THE COSMIC TOURIST

THE
COSMIC TOURIST

THE 100 MOST AWE-INSPIRING
PLACES IN THE UNIVERSE

CARLTON
BOOKS

A CIP record of this book is available from
the British Library.

ISBN: 978 1 78097 837 6

Created by Canopus Publishing Limited
www.canopusbooks.com

For Canopus:
Director/Editor: Robin Rees
Design/Art/Technical: James Symonds
Copy Editing: Sarah Tremlett
Proof printouts: Minuteman, Bristol
Portrait photography: James Symonds

For Carlton
Publishing Director: Piers Murray Hill
Executive Editor: Gemma Maclagan Ram
Art Director: Clare Baggaley
Proofreader: Barry Goodman
Production Manager: Maria Petalidou

Image opposite: The Soyuz TMA-01M
approaches the International Space Station.

Image on following page: Spitzer Space
Telescope mosaic of our Milky Way Galaxy.
The stars crowding the centre of our Galaxy
create the blue haze, and the brightest
white feature is the central star cluster in
our Galaxy. The green features are star-
forming regions. A massive black hole lies

ITINERARY

PREFACE

Once upon a time there was *Bang! The Complete History of the Universe.* It was written over a period of about two years, by the three of us – each a very different sort of astronomer. It was dedicated to telling the complete story of the creation of our Universe in the strict order in which it happened, in a way which could be understood by anyone who had an appetite for this fascinating new science. The book has now run to four editions in 15 languages and is proving highly popular. Popular it should be, because the current wave of new discoveries about the Universe in which we live affects all of us – and shapes the way we see ourselves in the grand scheme of things.

Encouraged by this success, we now present a new labour of love ... *The Cosmic Tourist.* This book has been written to a new discipline – it is a journey which begins on Earth, and takes us ever outwards, further from our birthplace, until we reach the very edge of the observable Universe. We describe the greatest of all possible tours – visiting one hundred of the most extraordinary locations currently known to mankind.

You can keep up to date with *The Cosmic Tourist* and *Bang!* at our website www.BangUniverse.com

Our special thanks to Sara Bricusse, Julia Knight, Ptolemy the feline escapologist (care of Sir Patrick), Derek Ward-Thompson, Neil Reading, Noah Petro, Iain Nicolson and Phil Murray.

Brian May
Patrick Moore
Chris Lintott
July 2012

Notes on units
1 light-year = 5,878,625,373,184 miles or 9,460,730,472,581 kilometres
1 light-hour = 670,616,628 miles or 1,079,252,850 kilometres
1 light-minute = 11,176,943.8 miles or 17,987,547.5 kilometres
1 light-second = 186,282 miles or 299,799 kilometres
1 billion = 1000 million or 1,000,000,000
1 trillion = a billion billion

Notes on classification
The brightest stars in the 88 recognized constellations have proper names, for example, 'Betelgeux' in Orion. These stars, and the fainter stars in the constellation are also classified using Greek letters – alpha, beta, gamma, etc. from brightest to faintest, followed by the constellation name. So Betelgeux is also known as Alpha Orionis.

Outside our Galaxy, remote galaxies and nebulae are usually identified using Messier (M) numbers, New Galactic Catalogue (NGC) numbers, or Caldwell (C) numbers, and these are given along with the names by which each object is more commonly known – for example the Andromeda Galaxy is designated M1. Where there is no common name for a stop on the tour, we just give the catalogue number.

BRIEFING NOTES

In this book we invite you to join us on a unique journey – to be a tourist on a scale never before imagined. Our plan is to take you to see one hundred of our favourite places throughout the entire Universe. Some are spectacular, some are intriguing, and some – well, we just like them. In any case, each has its own particular tale to tell. We will begin near home, and gradually work our way out to the extreme limits of what is currently known.

How can we travel to these remote places? Even if we travel at the maximum speed allowed by the laws of physics as we know them – at the speed of light – we still will not get very far in a human lifetime (though Einstein's General Theory of Relativity predicts that at speeds very close to light speed, an astronaut might age very little in a round trip that, to historians on Earth, spanned hundreds or thousands of years – the inspiration for the song *'39* written by BM). Luckily, for the purposes of this book we have access to a rather special spaceship, Ptolemy, named after Patrick's cat, of course. This ship will simplify the whole thing; it travels at the speed of *thought*. We can set the coordinates of any cosmic scenic location on its control panel, and instantly Ptolemy will take us there, so we can see it with our own eyes.

Keeping track of our position on the way will need some thought. Even in the Solar System normal earthly units of measurement are clumsy, to say the least; measuring the distance between the Earth and the Sun, let alone the size of our Galaxy, in miles makes as little sense as measuring the distance between London and Moscow in inches.

For a cosmic measuring stick, astronomers have adopted light itself. Light travels at 186,000 miles per second – or about 300,000,000 metres per second – in space. This is relatively slow compared to Ptolemy's top speed, but still, to us humans, almost unimaginably fast. We can measure our distance from home in terms of how long it would have taken us to get there at the speed of light. Visiting one of the planets in our own Solar System, it would take us only a few seconds to get there at the speed of light – so we say we are just a few light-seconds away. But somewhere out among the Sun's neighbouring stars, it might have taken light a hundred years to arrive; so in this case we would say that we are one hundred light-years away. This whole business of light taking a finite time to get from place to place plays tricks with our minds, because we never experience it in our daily lives. But most of us are now quite used to the rather odd concept that the light we see from the stars and other objects in the heavens left those stars a long time ago – in some cases thousands of years in the past. So the further an object is from us, the further back we are effectively looking in time. Amazingly, looking up into a dark night sky, we see the famous Andromeda Galaxy (even with the naked eye, if we are lucky) as it was 2.6 million years ago. It is very possible that some of its features are actually quite different right now from what we see – but ordinarily we would not find out for the next 2.6 million years. Our spaceship Ptolemy, on the other hand, can move instantaneously around in space, and is able to show us what every cosmic tourist attraction is like at this very moment!

Without wishing to spoil any of the surprises, it's perhaps as well to keep in mind that no matter how far away we journey on this adventure, it will still be the twenty-first century. And when the tour is completed, we will be able to confirm that wherever we wander, the Universe, on a broad scale, will always look very much the same. This is one of the basic beliefs currently held in cosmology ... that there is no centre to the Universe – everything we can see was once contained in an infinitesimally small point when the Big Bang began, but that point is now everywhere, since space itself expanded along with the physical Universe. However, moving through observable space, we will find a vast array of different wonders, more astounding than could have been created in the imagination of the finest science-fiction writer. The Universe is indeed an unending source of breathtaking vistas, and mind-blowing concepts, and no Cosmic Tourist will ever forget the trip.

We have a long journey ahead, your pilots are ready, and there is plenty to see.

Ptolemy is well equipped for a cosmic voyage; not only will it keep us safe from harmful radiation and extreme cold in the vastness of space, and from being crushed by powerful gravitational fields, but also, by means of special sensors and filters, it will allow us to see far more than we could with our eyes alone.

Fasten your seat belts please.

Enjoy!

PLANET EARTH

Distance from Earth: 0

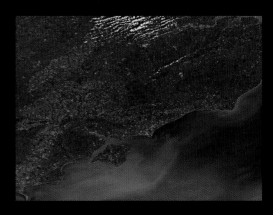

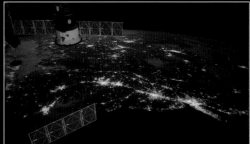

Among all the interesting objects we will encounter in our tour of the Universe, the Earth is the only one we already have first-hand, intimate knowledge of; after all, our feet are in contact with it for much of our lives.

However, to see Earth in context, in its full glory as a planet, we have to go into space. So, let us board our spaceship Ptolemy for the first leg of our cosmic journey – a very short hop – to orbit alongside the International Space Station (ISS), about 250 miles (400 kilometres) above sea level. From this position planet Earth is a magnificent sight; the view is dominated by the blue of the oceans, contrasting with the brown and green of the continents and the white of the ice caps – the whole laced with ever-shifting swirls of cloud.

It seems strange to realize that early civilizations thought that the Earth was flat, and that only a few centuries ago it was still believed to lie in the centre of the Universe, with the entire sky revolving around it; we know better now. Our world is a planet subject to changes, which are often unpredictable and it can be violent and uncontrollable; we cannot regulate the weather or subdue a tsunami. From space, though, we can see the Earth much more clearly, whether to follow the path of a storm, or to keep an eye on the spread of vegetation.

From our usual surface-bound perspective, we know that planet Earth is teeming with life, but is there any sign of it from orbit? What

ABOVE: A new whole Earth image captured by the Suomi NPP satellite in 2012.

OPPOSITE TOP: Sir Patrick Moore lives on Selsey Bill, which protrudes into the English Channel, situated just to the right of the island (Isle of Wight).

OPPOSITE MIDDLE: The Eastern Seaboard of the United States from the Washington DC area up to Rhode Island.

BELOW: Sunset over the Pacific Ocean seen from the ISS. The anvil tops of thunderclouds can be seen.

would aliens stopping off on their own cosmic tour conclude about this small world? Perhaps the most striking clue is the way that the night-time Earth sparkles, lit up by the glow of millions of artificial lamps. We can imagine the arguments of the alien tourists as they seek to interpret these strange illuminations; could they be natural? The lights seem arranged in an intricate pattern, which might suggest an artificial origin, but clinching evidence is not far away.

From our position in orbit, with just a small pair of binoculars we can pick out signs of civilization; cities, roads, and even ships' wakes, carved out in the oceans, are easily visible. The atmosphere, too, betrays the existence of life; the presence of significant amounts of oxygen is sustained only by the efforts of plants, which constantly replenish it. Earthbound observers are perhaps just a decade away from being able to detect life on distant worlds in just this fashion, and so, viewed from orbit, there can be no doubt that the Earth is inhabited. Whether the human race counts as intelligent life or not, we're not sure!

In any case, we are not alone in space; in fact, it's pretty untidy around here. Our vantage point is shared with over half a million items of man-made debris; around one thousand are working satellites, but most are space junk, tiny pieces of metal from previous missions. Some of the pieces are the results of collisions; others are the remains of satellites destroyed in weapons testing, while the remainder are sections of old rocket boosters, which exploded after use. The most popular orbits are becoming crowded, and satellites now have to swerve to avoid collisions.

With all of this activity it's hard to remember that the Space Age only began in 1957, with the first artificial satellite, Sputnik 1. Attempts had been made to reach space before. There is a story, apparently fairly well authenticated, that in 1600 a Chinese enthusiast named Wan-Hood tried to launch himself into space by fixing forty-seven gunpowder rockets to his body and ordering his servants to light them simultaneously. The experiment was not a success, but at least he didn't have to worry about space junk.

THE MERRY DANCERS

Distance from Earth: 0.0003 light-seconds

Our orbit carries us quickly over the surface of the Earth; at this altitude we complete one orbit roughly every ninety minutes. As we cross from day into night, we might well notice a flickering light, high in the atmosphere, around the poles. In fact, both the North and South Pole are surrounded by a complete ring of light, which at its brightest has a distinctively green tint.

These are the aurorae – the lovely polar lights – which would look quite different to observers down on the ground. From Earth an auroral display appears not as a complete ring, but as a glow toward the horizon, an arc above it or, for the lucky observer, as dramatic flickering curtains. From a dark site, the Northern Lights can be an eerie sight, and perhaps it is not surprising that there are many legends about them. In Scotland the lights were known as 'Merry Dancers'; in Finnish lore they were the 'Revontulet' or 'fox fires' – sparks whisked upward by the swishing tails of foxes which lived in Lapland, and were made of fire; while in Old Norse legend the lights were reflections of the clouds stirred up by 'swarms of luminous herring' swimming in the Arctic Ocean.

The true explanation of the Northern Lights is no less fascinating. Earth lies not in empty space, but amidst a stream of energetic particles thrown out into the Solar System by the Sun itself. As they approach the Earth, most of the particles are deflected by our magnetic field, but others are caught and channelled down towards the poles, where they collide with the molecules which make up the tenuous gas of the upper atmosphere, exciting them and making them glow. This process forms the Aurora Borealis in the northern hemisphere, and the Aurora Australis in the southern. It is the shape of the Earth's magnetic field which guides these particles, forming the auroral glow into the rings we see from space.

The solar wind does not blow with a constant strength; it is often gusty! When the Sun is active and the solar wind is strong, then the ring of light may reach down toward the equator allowing the Aurora Borealis to be seen from southern Europe, and very occasionally, even as far south as Singapore, which witnessed a dramatic display in 1908. Without waiting for such spectacular events, getting closer to the poles increases the chance of a good display – but don't get too close. The poles themselves, which lie not under, but at the centre of the auroral ring, miss out on the best views.

Spectacular, yet elusive, the aurorae of Earth are unforgettable – even for a tourist travelling as far as we will be. Before we leave the vicinity of our home in space, we must pay tribute to a remarkable satellite, also orbiting in a 'low Earth orbit' – the Hubble Space Telescope.

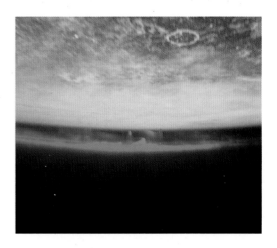

ABOVE: The Aurora Borealis seen from the International Space Station (ISS). The circular feature is the Manicouagan Crater in Canada.

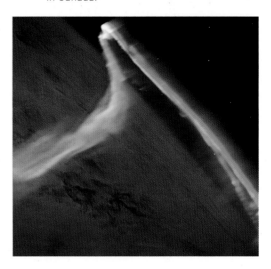

ABOVE: The Aurora Australis seen from the ISS above the Indian Ocean.

OPPOSITE: The Aurora Borealis over the Midwest of the United States, taken from on board the ISS.

OUR EYE IN SPACE

Distance from Earth: 0.002 light-seconds

Of all of the man-made satellites, one holds a special place in the heart of astronomers. Way back before the Space Age had even begun, astronomers had dreamt of putting a telescope up above the Earth's atmosphere, escaping the infernal twinkling that makes observing from the surface so hard. The Hubble Space Telescope, launched in 1990, is the fulfilment of this dream, and it has been an enormous success; it has a good claim to be the most productive scientific instrument in history.

It was not always plain sailing, however. Shortly after Hubble's deployment in orbit around the Earth, it was realized that the images were blurry – the result of a missing washer in a piece of optical testing equipment that had been used to prepare the telescope's primary mirror. Luckily, Hubble had always been designed to be serviced, and a repair mission making use of the Space Shuttle's ability to capture and hold satellites was quickly organized. A set of complicated 'glasses', known as COSTAR, was fitted to the telescope, and the problem was solved. Further Hubble repair missions followed, as astronauts replaced instruments, added a new camera and even repaired computer components buried deep within the telescope's structure. These bravura displays of sheer technical excellence were some of the manned space programme's greatest achievements, and the astronauts who carried them out are heroes to the many thousands of astronomers who have taken advantage of their handiwork; but those days are over. The retirement of the shuttles, in 2011, means that no further missions can take place. The last crew to visit Hubble added a docking station for a future robotic mission which will, sometime in the next few years, after Hubble has ceased making observations, drag it down into the Earth's atmosphere to burn up as a spectacular meteor. Hubble is large enough that it is thought to pose a danger if left to decay on its own; nonetheless, such a wasteful end for such a marvellous telescope seems undignified, to say the least.

For now, though, Hubble shines, its panels reflecting the sunlight that powers it, as it sails high above the Earth's atmosphere. Rather than detailing the telescope's scientific legacy here, suffice it to say that there is scarcely a topic in this book that hasn't been touched by observations carried out by this amazing instrument.

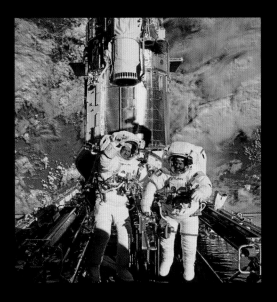

ABOVE: Astronauts John Grunsfeld and Richard Linnehan with the Hubble Space Telescope in Columbia's cargo bay.

ABOVE: Fixing the Hubble Space Telescope. Astronaut Michael Good can be seen attached to the shuttle's robotic arm.

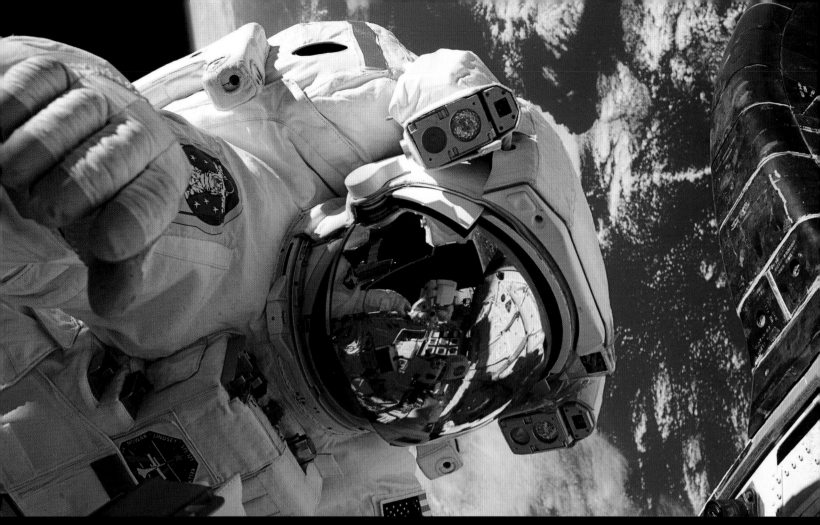

ABOVE: Piers Sellers spacewalks from the Space Shuttle Discovery. Photographer Michael Fossum is reflected in his visor. Without spacewalks like this, many of the Hubble's discoveries would never have happened.

BELOW: The Hubble Space Telescope floating in space, seen from the departing Space Shuttle. The solar panels are retracted in this shot.

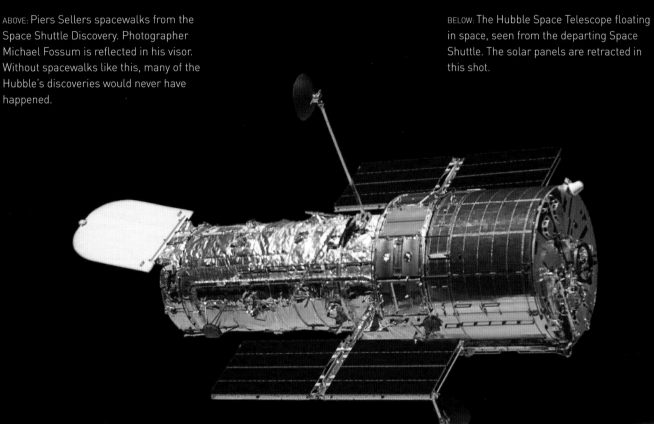

HOW HIGH THE MOON

Distance from Earth: 1.28 light-seconds

We now turn Ptolemy outwards, and head into space toward a new target – the Moon. As long as humans have walked the Earth, they have been fascinated by the silvery disc in the sky, whose shape waxes and wanes every month, and whose 'face' tantalisingly suggests human features. The Moon plays a part in the legends of every culture and has been a symbol of romance to many; but we are about to see it close-up, in stark detail.

The Moon is our planet's 'little sister', its closest companion, and, while the 250,000 miles (400,000 kilometres) or so that separate it from Earth may not be far on a cosmic scale, it is a daunting enough journey for mankind. A beam of light can make it from the Earth to the Moon in just over a second, but the Apollo astronauts, propelled toward the lunar surface by the most powerful rocket ever constructed – the Saturn V – took three days to get there. Their spacecraft was configured so that they sat with their backs to the Moon for much of the way, so they were only able to turn to see their destination as they neared lunar orbit.

We, in Ptolemy, will do things differently, allowing the Moon to fill our view from the portholes as we approach it. Even from Earth the principal features that cover the surface are easily visible, but as we draw closer we can see them in ever greater detail. The Moon is a world of mountains, valleys, and craters, with dramatic landscapes and impressive scenery. Unlike the Earth we have left behind, it is not a colourful world. In the absence of water, air and life, there is a certain bleakness to the scene; Buzz Aldrin of Apollo 11 summed it up in two words: 'Magnificent Desolation'.

Before landing, let us get a global view. The most obvious features are vast dark areas which have traditionally been called 'seas'; in fact they have nothing to do with water, but were formed by lava which spread across the surface billions of years ago. These lava lakes cooled and solidified, and, apart from the effects of relatively minor meteorite impacts, have changed little since then. As a result, the lakes are flatter than their surroundings and thus provided a safe landing site for all but one of the Apollo missions.

Craters are everywhere, of all sizes; they cluster thickly on the uplands, and even the 'seas' are not completely free of them. Some have terraced walls, and many have central mountains or mountain groups; a few craters are the centres of systems of bright rays which extend for hundreds of miles in all directions. Seen from Earth, two particular ray systems dominate the entire scene around the time of full Moon. One of these ray-centres, in the southern uplands, is called Tycho, named after a famous sixteenth-century Danish astronomer,

ABOVE: The Apollo 11 Command Module in orbit above the surface of the Moon.

ABOVE: Moonrise, seen from the Space Shuttle Columbia.

ABOVE: The white lunar crater at left is Copernicus, and the pair of nearly circular dark features to the right are the Mare Serenitatis above and Tranquillitatis below.

Tycho Brahe. It is over 50 miles (80 kilometres) in diameter, so let us bring Ptolemy down above it. We find the ray-centre is impressive with high walls and a sunken floor, but where are the rays? Since they are composed of nothing but thin surface deposits, casting no shadows, the ray-centres are hard to see in close-up, unless the sunlight catches them at a particular angle.

We see that the craters are normally circular. Seen from Earth all craters near to the edge of the disc appear foreshortened into ellipses, but if we guide Ptolemy above a crater such as Plato, its circular form, too, is seen to be perfect. As we pass Plato we have an excellent view of the Lunar Alps, cut by a broad valley 80 miles (128 kilometres) long.

What made the craters? They look rather like old volcanoes – but they are not; they were produced by the impact of meteorites, and they are very old. The entire surface was moulded by what we call the 'Great Bombardment' ending around three billion years ago, when the last phases of the formation of the stable Solar System we see today led to the inner planets being peppered with asteroids and comets. The record of this violent time has long been erased on Earth, where weather and the movement of tectonic plates do not let anything last for long, but the Moon still bears the scars. Its pock-marked surface bears witness to the Solar System's history, but relatively little has happened to the Moon for millions, if not billions of years. If the dinosaurs looked up at the Moon, they would have seen it very much as we do today.

TRANQUILLITY BASE

Distance from Earth: 1.28 light-seconds

Even tourists who cannot, like us, travel at the speed of thought, may one day be able to visit this location – surely a prime tourist destination of the future. This is Tranquillity Base, where Neil Armstrong's 'one small step' confirmed that mankind was capable of leaving our home planet and travelling to a neighbouring world.

There isn't much colour to the bright lunar surface, true, but the landscape is spectacular and the contrast between it and the deep, dark blackness of space is perpetually breathtaking. As Armstrong exclaimed a few minutes after disembarking, the Moon has 'a stark beauty all its own'.

Even from orbit we can easily see the evidence of the Apollo 11 astronauts' visit. The lander itself is visible, shining brightly in the sunlight, and the tracks marking the small excursions made by the crew are visible too. Images of the moonwalk from the mission show a small crater about 165 feet (50 metres) from the Eagle, and we can see the track lead up to it and double back. A few items of discarded equipment are visible, some of which were left behind to continue monitoring conditions after the astronauts had left, along with the disturbed region marked by the ascent of the Eagle's upper stage.

Left alone, these artefacts will endure for eons on the lunar surface, along with the famous flag and the plaque bearing the reminder that Apollo 11 'came in peace for all mankind'. Without the Earth's atmosphere to weather them away, the footprints of Armstrong and Aldrin will probably last for thousands of years, inspiring subsequent generations of travellers with the tale of the pioneers who risked everything to spend a few hours on the Moon.

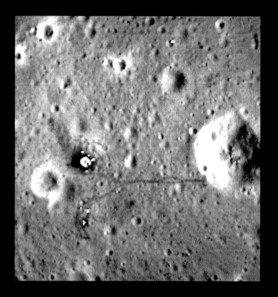

ABOVE: Detail of the area around Apollo 11's Lunar Module, with the experimental equipment to the south and the tracks left by the astronauts leading to the crater.

OPPOSITE TOP: On 20 June 1969 the Apollo 11 astronauts took off from the Moon, leaving the descent stage of the Eagle, Apollo 11's lunar module, behind them on the surface. More than 40 years later, this Lunar Reconnaissance Orbiter image shows the Eagle standing amongst the craters of the Sea of Tranquillity.

OPPOSITE BOTTOM: James Irwin salutes the flag during the Apollo 15 mission.

LEFT: Earth seen from the Apollo 11 landing site with the Lunar Module in the foreground.

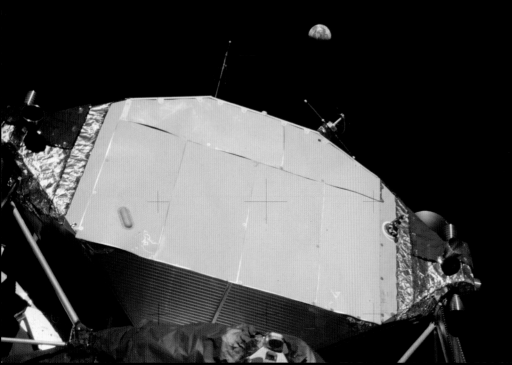

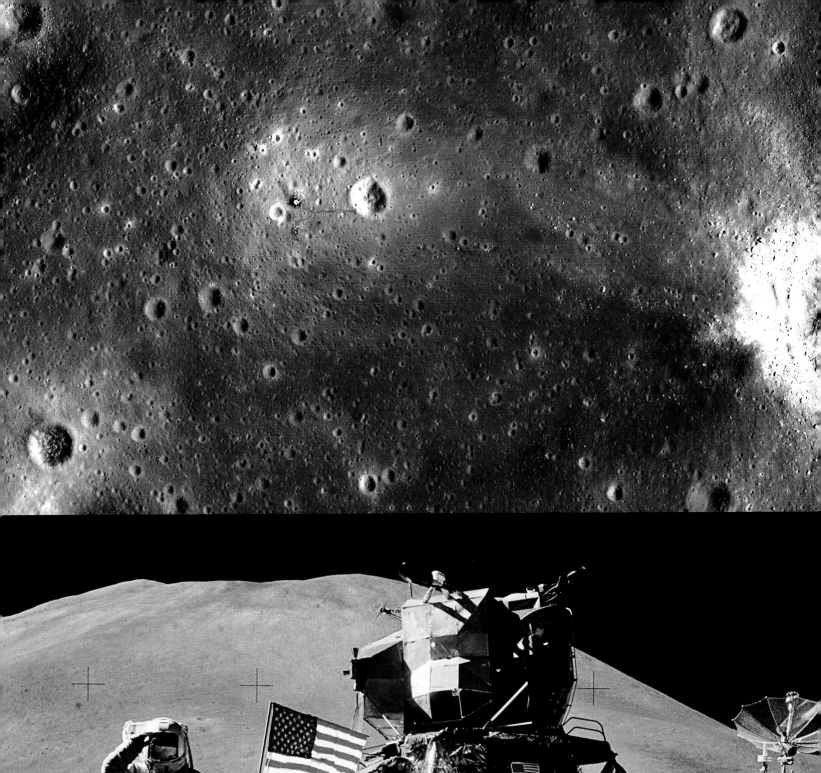

THE STRAIGHT WALL

Distance from Earth: 1.28 light-seconds

Our next stop on the lunar surface is sure to be one of the Moon's most popular natural tourist attractions if, some day, travel here becomes commonplace. One day people may flock here just as they do to Earth's Grand Canyon, but for now we will have this magnificent view to ourselves. To navigate to it, we must first find the Ptolemaeus chain of craters, near the centre of the face of the Moon that's always turned toward the Earth. Ptolemaeus – named after one of the greatest Greek astronomers – is the largest member, 92 miles (148 kilometres) in diameter, with walls that rise just a mile and a half (2.4 kilometres) above the crater floor. Ptolemaeus is shaped more like a shallow pizza dish or tray than a deep bowl.

Distances on the Moon are deceptive. The Moon is a much smaller world than the Earth, so the curvature is greater and the horizon is only roughly half as far away as it is back at home. Standing in the middle of the crater is therefore something of a disappointment; we will not see the walls at all, because they will be below our horizon. This effect can make it hard to navigate, but from the perspective of Ptolemy we can find our way south, passing over the craters Alphonsus, then Arzachel and its neighbour Thebit, before reaching our quarry – a curious feature known as the Straight Wall.

The name is a little misleading, although it appeared as a straight line in the telescope of its discoverer, Johann Schröter, the first of the really great lunar observers. Nor is it really a wall; it is a fault in the surface which stretches for over 70 miles (112 kilometres) from its origin in a little clump of peaks, known as the Stag's Horn Mountains. As we approach we move just beyond the edge of the Mare Nubium, or Sea of Clouds and then suddenly, as we reach the fault, the ground to the west drops by a full thousand feet.

The dramatic view as observed from Earth at the right time of the month, before full-on sunlight comes from the east, is due to the black shadow that this escarpment casts on the plains. After full Moon the sunlight comes from the opposite direction, and shines on the inclined face and illuminates it so the 'Wall' shows up as a bright line.

Viewed from the lunar surface, it's the Wall's immense length that impresses. Travellers expecting a sheer cliff will be disappointed, as the angle of slope is nowhere more than 40 degrees. It's still impressive when the light is right, though, and standing at the edge of the shadow allows us to demonstrate a favourite trick of the lunar tourist – making yourself disappear.

On Earth, shadows are never really pitch black. Light which scatters as it passes through the Earth's atmosphere lights up even the darkest part of the shadow, but the Moon has no atmosphere and so no light scatters. Anything which passes into a lunar shadow is almost completely lost from view – and where better to play lunar hide and seek than around the Straight Wall?

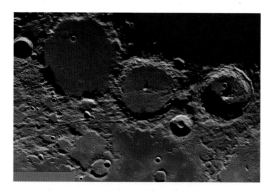

ABOVE: The chain of three large craters: Ptolemaeus, Alphonsus and Arzachel.

ABOVE: The crater Ptolemaeus photographed from Apollo 16.

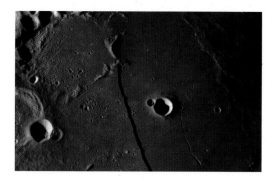

ABOVE: The Straight Wall photographed from Earth by Damian Peach.

WATER ON THE MOON

Distance from Earth: 1.28 light-seconds

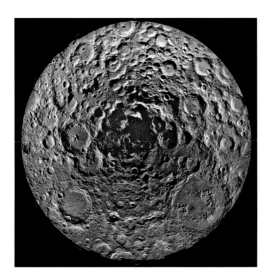

ABOVE: This mosaic from the satellite Clementine is dominated by the major depression centred near the South Pole.

ABOVE: Cabaeus crater, target for the LCROSS mission which searched for water in its dark recesses.

We now travel southwards, moving quickly above the lunar surface. The Moon is, after all, a small world, only a quarter of the diameter of our own planet, and soon we are treated to a fine view of regions that are hard to see from the Earth. Looking down, we can see the vast South Pole-Aitken Basin, one of the largest impact structures in the Solar System. From Earth, it appears right on the edge of the Moon's disc, and so is foreshortened, but Ptolemy can take us right into it.

The basin is vast; covering an area more than eight times that of the British Isles, it is 1500 miles (2400 kilometres) across and more than eight miles (13 kilometres) deep. So we have to rise to a few miles above the surface to get a sense of it as a whole. It seems that this huge basin formed as a result of a dramatic impact early in the Moon's history. Although such massive collisions are extremely rare now, they must have been relatively common during the time of the Great Bombardment when it was impacted by many asteroids and comets.

From this range we can also see interesting features within the basin; its interior is filled with a myriad smaller craters.

One of these craters is Cabaeus, which was the target for the daring 2009 LCROSS (Lunar CRater Observation and Sensing Satellite) mission. The goal of LCROSS was to search for water on the Moon, a valuable potential resource for future manned missions. This might seem like the ultimate quixotic quest – it has long been known that liquid water simply cannot survive on the airless lunar surface. The 'seas' are nothing more than lava plains, and anyone hoping to go for a swim in the Bay of Rainbows is certain to be disappointed. However, scientists suspected that down under the dark floor of Cabaeus, things might be very different.

The upper stage of the parent rocket that took LCROSS to the Moon crashed into the crater floor, ahead of the tiny spacecraft itself; this produced a large impact, throwing up debris which LCROSS was able to analyse in the few minutes before it, too, crashed into the crater floor. The analysis indicated the presence of substantial deposits of water amongst the material thrown up by the impact; more than enough, in fact, to keep future astronauts going.

Where did this water come from? It may have been produced locally, through the subtle interactions of the lunar surface with particles that flow from the Sun. Another more dramatic theory, is that the water was brought to the Moon by the impact of comets.

This water apparently survives because the floor of Cabaeus, shielded from the Sun's light by the crater walls, is one of the coldest places in the Solar System. Temperatures in nearby craters have been measured at −248 degrees Celsius – more than cold enough to keep deposits of water permanently locked up in the ground.

One day, astronauts may mine the bottom of craters like Cabaeus for their water.

MONARCH OF THE MOON

Distance from Earth: 1.28 light-seconds

We will now withdraw to several thousands of feet above the actual surface, so that we can take a broader view of the features of the Moon. We will aim to visit Copernicus, in the Ocean Procellarum. Its name was given by an astronomer named Giovanni Battista Riccioli in 1651, when he drew the first very useful lunar map; and there is a story here. Riccioli believed the Earth was at the centre of the Solar System – the conventional theory, Ptolemy's theory, almost universally believed at the time. Copernicus had suggested that the planets moved around the Sun, a viewpoint which turned out to be correct, but was then viewed as heretical. And so, when Riccioli named the craters, he scornfully 'flung Copernicus into the Ocean of Storms'. If his plan was to scorn his rival, the plan misfired; the crater he selected is one of the most impressive on the entire lunar surface, and has been named 'the Monarch of the Moon'. It is one of the two great ray-centres (Tycho is the other).

If we bring Ptolemy down to only a mile (1.6 kilometres) or so above the surface inside the crater, we can appreciate the high, beautifully terraced walls, which rise to well over two miles (3.2 kilometres) above the sunken floor. Because Copernicus is relatively young, it has not been flooded by lava, but a very different fate has overtaken Stadius which lies nearby to the east. We lift ourselves out of Copernicus across the wall, and move out on to the plain in the west to see Stadius – but it is a 'ghost'. It may once have been deep and imposing, but Stadius has been so drowned by lava that its walls are barely traceable. Ghost craters are not uncommon on the Moon, but Stadius is a particularly good example.

Let us look a little more carefully at the rays of Copernicus; they spread out in all directions, across the plains for 500 miles (800 kilometres). Using Ptolemy, let us follow one of the rays; it crosses all other formations, so Copernicus must have been formed at the very last stage of the Moon's active period, perhaps no more than 800 million years ago. The rays from Copernicus even overlie rays coming from another crater, Kepler. From Ptolemy we have a splendid view, but the rays don't cast any shadows; they are merely surface deposits formed at the time of the impact that produced the craters.

We are used to looking at the Moon from our vantage point on Earth, and the rays in Copernicus and the other great ray system, Tycho, dominate the entire lunar scene. But, since the rays were formed, the chaos has been replaced by total calm. In the near future there may be another change. Men from Earth may bring life back to the silent, waiting Moon, with flourishing lunar bases. Let's hope they keep the place tidy.

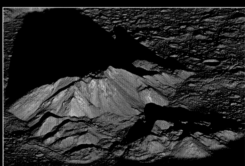

ABOVE: (upper) Apollo 12 image of the crater Kepler. (lower) Tycho's central peak casts long shadows near local sunrise, seen from the Lunar Reconnaissance Orbiter.

BELOW: In preparation for the Apollo landings, five Lunar Orbiter spacecraft were launched during 1966 and 1967 to gather information about the lunar surface. This dramatic mosaic of Copernicus was one of the outstanding results of these missions.

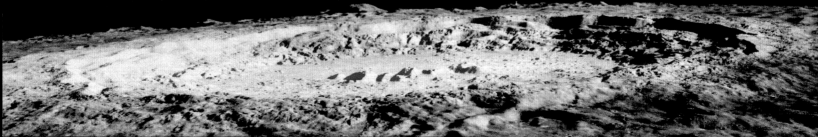

ARISTARCHUS THE BRILLIANT

Distance from Earth: 1.28 light-seconds

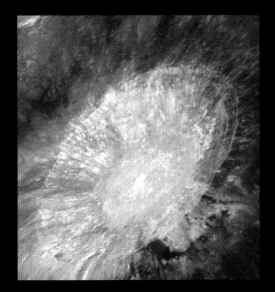

ABOVE: The Hubble Space Telescope's Advanced Camera for Surveys view of Aristarchus Crater.

BELOW: Aristarchus Crater's west wall seen by the Lunar Reconnaissance Orbiter from a distance of only 16.2 miles (25.9 kilometres).

The crater Aristarchus is by far the brightest feature on the lunar surface; it is not too distant from Copernicus, and our faithful Ptolemy can take us there instantly. As we approach we see the crater of Aristarchus, some 23 miles (37 kilometres) across, lying atop a rocky plateau in the grey expanse of the Ocean of Storms. Aristarchus was the intended destination of Apollo 18 but, because the Apollo programme ended in 1972 for various reasons, including cost, with number 17 the proposed explorations didn't happen.

From above, we see that Aristarchus has terraced walls which rise to well over two miles (3.2 kilometres) from the floor, on which there is a steep central peak. The floor shows streaks that are so unusual that there were once suggestions they could be due to lowly vegetation; only during the past few decades has it become clear that the Moon has never known life of any kind.

We can land quite safely inside Aristarchus, and pause to take in the scene. One point has to be borne in mind – the Moon doesn't have any air to burn up incoming meteoroids (the process by which we get shooting stars on Earth), so is there any danger of our being hit by small particles coming from space? It is true that tiny particles known as micrometeorites do bombard the surface constantly, but it's unlikely they will harm us. Over sufficiently long periods they tend to darken the surface, and the fact that Aristarchus is so bright shows that, by lunar standards, it is young – comparable in age to Copernicus and indeed may be even younger – perhaps no more than 500 million years. Just south of Aristarchus there is another crater called Herodotus, which is about the same size, but much less brilliant; let us bring Ptolemy right over it. Extending from Herodotus is a huge, complex valley, named in honour of the first great lunar cartographer Johann Schröter. There is nothing quite like it anywhere else on the Moon, and Ptolemy can take us right down to the floor, which is more imposing than being inside the crater, even though there is no real feeling of being shut in.

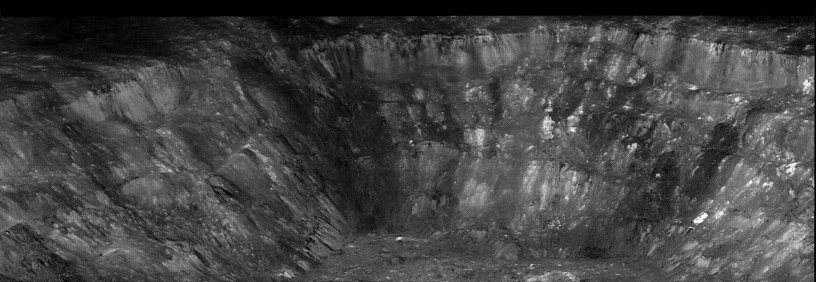

THE BAY OF RAINBOWS

Distance from Earth: 1.28 light-seconds

We really must pay a visit to the lunar landscape widely believed to be the most beautiful of all – Sinus Iridum, the Bay of Rainbows. This feature is always easy to find whenever it is sunlit; it extends out of the Mare Imbrium and is quite unmistakable. Sinus Iridum does take the form of a bay, but it is really a 150-mile (240-kilometre) diameter crater whose seaward wall has been more or less obliterated. The ends of the remaining parts of the wall are marked by Cape Heraclides and Cape Leplace. The site of the destroyed wall can just about be traced by some low, irregular ridges and a few tiny craterlets.

We can cast our minds back to views of Sinus Iridum as seen from the Earth. The bay is bordered by the Jura Mountains, beyond which comes the western part of the long, irregular Mare Frigoris (Sea of Cold); the ideal time to study this area is when the Sun has just started to rise over it. At that time of month the floor of the Bay of Rainbows is in shadow, but the tops of the Jura peaks are already illuminated, and seem to be separate from the Moon itself, producing an effect that is popularly known as the 'Jewelled Handle'. It makes a glorious photograph, but does not last long, and of course the Jewelled Handle is seen only once in every lunar day.

Let us take Ptolemy to a greater height above the surface, well to the east of Sinus Iridum. We pass over the dark-floored, 60-mile (96-kilometre) diameter crater Plato, and come to another remarkable feature, the Straight Ridge. It is 55 miles (88.5 kilometres) long, and even though the peaks are of modest height, they are arranged in such a regular way they look almost man-made (predictably some of the early lunar observers maintained that they were artificial).

ABOVE: Sinus Iridum is the large crater at right with the Jura Montes casting shadows in this image taken at sunrise.

BELOW: Sinus Iridum, the Bay of Rainbows is at left, ringed by the Jura Mountains. The crater Plato is at right.

THE DARK SIDE OF THE MOON

Distance from Earth: 1.28 light-seconds

ABOVE: The Lunar Reconnaissance Orbiter's image of the far side of the Moon is a composite of 15,000 images.

ABOVE: The diameter of the outer ring of the Mare Orientale is 528 miles (950 kilometres) seen in this mosaic from the Lunar Reconnaissance Orbiter.

So long as we stay on the Earth, there is part of the Moon that we can never see, the result of a tug of war between the two bodies that has been taking place for billions of years. The Earth is 81 times as massive as the Moon, but our small neighbour can still have noticeable effects. We know that the Moon is the cause of our ocean tides, or at least the main cause, and clearly the Earth could also produce tides in the Moon. The Moon today is solid, but in its young days it was viscous and subject to tidal distortion. At that early point in its history it spun round in a few hours, but tidal forces slowed it down, and today the 'day' on the Moon is equal to just over 27 of our Earth days. This means that the Moon goes round the Earth in exactly the same time that it takes to spin on its axis, and it always keeps the same hemisphere turned towards us. There is a slight 'wobbling', but 41 per cent of the Moon is never visible from Earth.

Before the Space Age we thus knew nothing about the Moon's hidden side, but, of course, both Russia and America sent rockets round to see what was there. Now, in Ptolemy we can make a much quicker survey of the 'other side of the Moon'. As we move past the edge of the Moon as seen from Earth, we come to a whole range of features which are slightly, but subtly, different. There is only one major 'Sea', the Mare Orientale (Eastern Sea), a huge ring structure, only a tiny part of which can be seen from Earth. One of us, PM, made an independent discovery of Mare Orientale in 1946. We fly along it, and admire the craters, pits and peaks; when we leave it and come to rest on the surface we find the usual craters and ridges, but an absence of very high mountain ranges. Our flights around the Moon have shown us a great deal, and we could stay for longer, but we are still in the early stages of our trip. Now we are really going to need Ptolemy, because we are going to venture to places no other kind of spaceship could possibly reach.

First, farewell to the Moon – we will ask Ptolemy to take us to that great beacon in the sky – the Sun.

It's tempting to rush headlong toward the Sun itself, but from a distance we actually get an excellent view of the Sun as a whole. In order to take a good look, Ptolemy will settle into a sight-seeing location between the Earth and the Sun, almost a million miles (1,600,000 kilometres) from the Earth. We are not alone out here; a posse of man-made satellites surrounds us, all pointing toward the Sun. This is a special point in space, known as the first Lagrangian point, L1. Here the gravitational pulls from the Earth and the Sun exactly cancel each other out, so it's a popular place for sungazing space vehicles to loiter, remaining always between the Earth and Sun without having to expend any fuel.

HERE COMES THE SUN

Distance from Earth: 8.3 light-minutes

Viewed from the Earth, the Sun is a brilliant yellow disc of light, usually speckled with a few dark spots (sunspots, which we shall investigate later). However, this apparent 'surface' – known as the photosphere – is not an indication of anything solid; the Sun is composed of gas alone, and the gases reach out millions of miles beyond the photosphere, into the space where the planets revolve. The reason we don't normally see this 'atmosphere' of the Sun is that the photosphere's brilliance overpowers everything else. We do get a glimpse of these outer regions from Earth during a solar eclipse of the

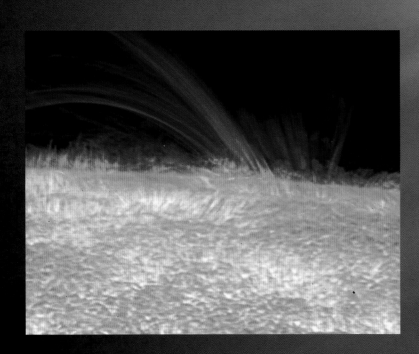

LEFT: The Hinode (Japanese for sunrise) Solar Optical Telescope revealed this fine structure in the chromosphere that extends outwards from the top of the convection cells, or granulation, of the photosphere.

BACKGROUND: Total eclipse, with superimposed photosphere showing sunspots.

Sun, when the Moon blots out the brilliant photosphere. Out here, in space, we do not need to wait for a total eclipse. We can simply mask the photosphere from our field of view, using a device called a coronagraph, and we can then see the beautiful spectacle of the Sun's atmosphere.

The lowest level of the Sun's atmosphere appears as a thin circle of reddish light. The deep red colour gives this layer its name – the chromosphere, from the Greek word for colour – and hints that the chromosphere, like the rest of the Sun, is almost entirely made of hydrogen, the lightest of the elements. This glowing layer is less than 2000 miles (3200 kilometres) deep, but it is home to the solar prominences – mighty eruptions of gas which rise, curl around under the influence of the Sun's magnetic fields, and fall back on to the glowing surface. These spectacular tongues of fire look small at this distance, but could swallow up our planet Earth many times over.

But there is more; the extended atmosphere of the Sun is known as the corona, a pearly, iridescent tracery of streamers of gas reaching out into the blackness, which here, from our position in Space, seem about to envelop us.

RIGHT: Sunset, as we see it on Earth, captured by Pete Lawrence in Selsey, England.

THE SUN VIEWED IN A DIFFERENT LIGHT

Distance from Earth: 8.3 light-minutes

Keeping our gaze fixed on the Sun, we can learn much more about it by using special, carefully defined colours of light to view it. We can, for instance, use a filter that only allows the deep red colour of the chromosphere through to our eyes, or our camera. This colour is the exact signature of radiation emitted by hot hydrogen; so, by using this technique, we get a fine view of the patterns that hydrogen gas is making on the Sun's surface. Several groups of sunspots that were too small to be seen using ordinary white light are now easily visible, together with a fine granular structure which covers the whole surface of the Sun.

We can adjust our filters to observe in other colours, as well, allowing us to focus our attention on the light emitted by other elements. By picking out light emitted by calcium, for example, we can study a thin layer of the chromosphere in which there are broad, disrupted regions, some much larger than the Earth, each surrounding a group of sunspots.

Given their great size, it's not surprising to discover that the spots are visible all the way back on Earth, but one must be very careful not to look directly at the Sun. To view the spots, you can either project the image on to a screen where it can be viewed in safety, or use strong, safe filters designed for the purpose. Of course, there is only one rule about looking at the Sun through a telescope – DON'T! Tragic accidents have happened in the past.

There's another problem with observing the Sun using our own eyes; they restrict us to observing only in visible light – the range of frequencies and colours to which our human eyes respond. But this is a tiny fraction of the total radiation emitted by the Sun. What we call

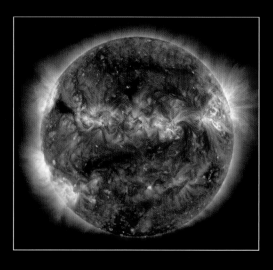

ABOVE: In this image of the Sun taken at three different wavelengths, blue is 1 million degrees, green is 1.5 million degrees, and red is 2 million degrees.

RIGHT: From left to right, this compilation of views of the Sun taken at different wavelengths comprises: the familiar optical view of the photosphere at 6000 degrees; the region between the chromosphere and corona at 1 million degrees (seen in extreme ultraviolet light where the active regions appear lighter); a composite of three different wavelengths showing temperatures up to 2 million degrees; an overlay of a science-based estimation of the magnetic field lines.

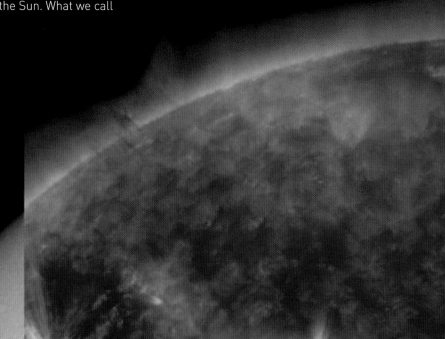

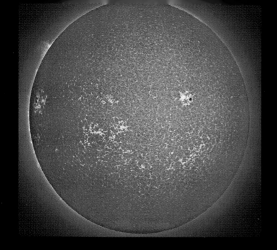

ABOVE: The Sun is not a perfect sphere. During periods of high activity the Sun develops a thin skin that increases its oblateness, or how much it is squashed at the poles. The glowing, white magnetic network, seen in this image taken using a violet calcium-potassium filter, is what gives the Sun its extra oblateness.

visible light is just a small part of the electromagnetic spectrum, which extends far outside what we can see. Light can be visualized as a wave moving through space, with red light having a longer wavelength than blue – corresponding to a lower frequency of vibration. Outside the visible range there is radiation with ever longer wavelengths – infrared, microwaves and radio waves, and on the other side, ever shorter wavelengths – ultraviolet, X-rays, gamma rays, etc.

Much of the Sun's output which actually reaches the Earth's surface is in the form of visible light, and this is the reason our eyes have evolved to be sensitive to this particular region; but the wavelengths outside this range tell us different parts of the Sun's story. The Sun's radio emissions, for example, rise and fall along with the general activity of the Sun. When it is active, with plenty of sunspots, then it is also 'bright' (or, perhaps, 'loud') in radio waves. Infrared observations show features in the chromosphere and corona silhouetted against a bright background, and a brightening at the North and South Poles, which are especially obvious when the Sun is quiet; indicating the presence of a 'coronal hole' – an area where the Sun's atmosphere is cooler than the average.

The really useful wavelengths for looking at the Sun, though, are those shorter than the wavelengths of visible light – which is the same thing as saying their frequencies are higher. These are ultraviolet (UV) and X-rays, and tuning into them shows us the hottest parts of the Sun's surface and atmosphere. The UV view reveals dramatic storms in sharp relief against a relatively dark disc, with great fiery arcs reaching up, and the corona clearly visible. We have no hope of obtaining this view from the ground, as Earth's atmosphere protects us from the harsh solar glare in the UV. From up here, though, it is obvious that these storms are linked both to the sunspot groups that are seen in visible light, and to the prominences.

In the domain of the still shorter wavelengths of X-rays, our view is dominated not by the solar disc itself, but by the corona. The hottest gas contained in the most active regions still shines brightly, but the corona is almost blindingly bright. The gas which it is composed of may not be very dense, but it is at an extremely high temperature; and, sitting at L1 (see page 29), we are almost inside it.

Let's now travel the remaining distance to the Sun, and take a close look at its inner workings.

STORMS ON THE SUN

Distance from Earth: 8.3 light-minutes

As we approach the Sun, the temperature of the outside of our spacecraft has been steadily rising. The corona is at a temperature of millions of degrees, but this does not mean that our ship is in danger of burning up – temperature and heat are not the same thing. Temperature is gauged by the rate at which atoms and molecules move; the faster they move, the higher the temperature. In the corona, the atoms are moving around very quickly indeed – hence the high temperature – but there are so few of them that little of this energy will be transferred to the outer shell of our ship, Ptolemy. A good analogy is that of a firework sparkler. Each spark of the firework is at a very high temperature, but there is so little mass in each spark that they can land on your hand quite safely. On the other hand, beware! Even after the sparkler is finished, the softly glowing end of the wire, though at a much lower temperature than the sparks, will burn you if you touch it, even for a moment, being of much greater mass.

How the corona got to be this hot is something of a mystery (it's much hotter than the surface of the Sun beneath it, which seems contrary to common sense); we don't understand all of the details yet, but the answer lies in the complex interactions of the Sun's magnetic field, and these have repercussions down on the Sun's surface, too.

For example, it's the magnetic field that's responsible for the dark sunspots. We can now manoeuvre Ptolemy so that we are directly above a sunspot – it's a relatively cool region amongst the bubbling mass of gas which is the Sun's surface. From our position poised above the spot it looks dark, but this is only in contrast with the brilliance of the photosphere around it. If the sunspot could be seen shining on its own, its surface would be brighter than an arc-lamp. The spot is not a simple patch; there is a dark, central portion which we call the umbra, around which is a lighter area called the penumbra. There is also a wealth of fine structure – streaks of dark and light – with bright filaments springing up above the sunspot, known as faculae (Latin for 'torches'). This particular sunspot is actually not on its own, but, as is often the case, part of a whole group of spots making a complicated pattern.

So, what exactly is a sunspot? The Sun has a magnetic field, and the lines of magnetic force run below the bright photosphere. There are some places where the field lines have become tangled, and have broken free on to the surface, moving material around, making cooler regions to produce such a group of sunspots. Were the Sun a solid body like the Earth, these lines of magnetic force would spin round with the Sun, and there would be little magnetic disturbance and no sunspots. However, as we have noted, the Sun is not solid, but

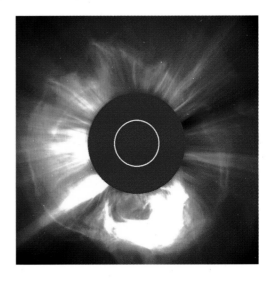

ABOVE: The cloud, or coronal mass ejection, in this image from SOHO is one of the strongest ever witnessed. It occurred on 28 October, 2003.

ABOVE: A highly detailed image of a sunspot from the Big Bear Solar Observatory, taken in visible light.

ABOVE: Massive coronal mass ejection seen by the Solar Dynamics Observatory.

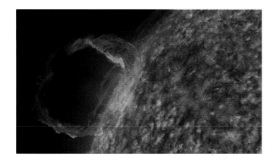

ABOVE: Solar Dynamics Observatory close up of a massive loop of gas.

ABOVE: In January 2012 one of the strongest solar storms for several years was witnessed by the Solar Dynamics Observatory.

composed of hot gas. The material near the equator is rotating faster than that near the poles, and this distorts the regular magnetic field. Its attempts to snap back to its resting position produce most of the 'weather' we see at the Sun's visible surface.

We are lucky to be able to see these sunspots, because there are times when there are none at all. The Sun has a definite cycle of activity, with a period of around 11 years. Every 11 years, at maximum activity, there may be numerous groups active at the same time. At solar minimum the Sun may be without spots for many days at a time. Because the processes which govern the appearance and disappearance of sunspots are complex, the cycle isn't entirely regular, and the most recent minimum was particularly pronounced and extended. Yet, even when the Sun looks quiet and calm, its surface is in a constant state of turmoil.

Ptolemy is able to give us a really good view, but our eyes are only safe because of the incredible filtering properties of its windows. The spots can be clearly seen, and if we watch them from day to day we will see that they are being slowly carried across the disc, by virtue of the Sun's rotation. The Sun takes about a month to spin once on its axis, and this means that it takes a fortnight for a spot to be carried from one side of the disc to the other. A fortnight later it will reappear at the opposite edge – provided that it still exists. Remember, the Sun is gaseous, and the sunspots are not permanent. A large group may persist for many weeks, but a small spot may be gone within an hour of being formed.

No spacecraft has ever been as close to the Sun as Ptolemy is now. The former record holder was Helios 2, which approached as close as 27 million miles (43 million kilometres) to the Sun. This German–American spacecraft (which survived many years in a harsh environment) still quietly orbits the Sun, even though nothing has been heard from it since the mid-1980s. The European Space Agency, ESA, is currently planning an ambitious follow-up mission, known as Solar Orbiter, which will orbit just 25 million miles (40 million kilometres) above the surface, providing spectacular imagery for as long as it can survive. However, we can do better than that; thanks to the advanced technology on board Ptolemy, we can plunge into the Sun itself.

The Sun's gravity dominates the Solar System; it keeps the Earth and the other planets in their orbits. Therefore it might seem that falling toward the Sun would be as simple as letting gravity pull us in, in the same way that hang-gliders descend gracefully under Earth's own relatively puny pull without the need for power. But actually, moving toward the Sun takes a lot of effort and the engines of Ptolemy have been working hard ever since we left the Earth and Moon behind.

The reason for this is a fundamental principle known as the conservation of angular momentum. The planets, and our spacecraft too, since it started out from an orbiting Earth, are not static in space, but are moving around the Sun. To move inwards we have to give up some of this angular momentum – which means braking – and that takes effort. Solar Orbiter will use Venus to do this work, swinging by the planet seven times in order to sacrifice momentum, but we can let Ptolemy's engines do the work.

THE HEART OF THE SUN

Distance from Earth: 8.3 light-minutes

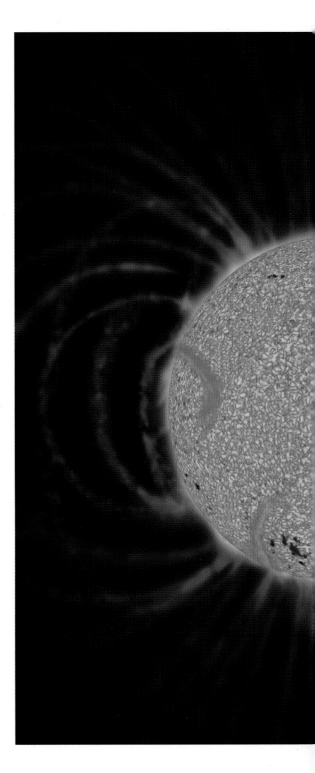

Heading toward the photosphere, we now get a better view of the processes which make the Sun's surface such an interesting place – so different from the simple yellow disc that features in children's paintings.

There is no solid surface to block our passage toward the centre of the Sun. As we descend, the density of the gas outside Ptolemy increases, and so does the brilliance of the glare and the temperature. We might still be visible to powerful telescopes back on Earth as a tiny sunspot, black against the brilliance of the Sun's disc, but as we sink through the chromosphere and then beneath the photosphere we will vanish from view.

This is not a calm place. The source of the Sun's energy lies a long way beneath us, and only a subtle and unstable balance between the pressure of this escaping energy and gravity prevents the Sun from collapsing. We pause in our descent for a second, and are immediately carried off by a powerful current. In this part of the Sun, transport of energy takes place by convection. Material from below us is heated by the Sun's core, and, as a result, rises – just as the hot air in a balloon causes it to rise through the Earth's atmosphere. As material rises it gives up its energy and cools, before sinking once again. The process – convection – is the same as that which drives major winds back home on Earth, where warm air rises at the equator, cools in the upper atmosphere and then sinks back down to the surface.

The current we're riding takes us a long way down, more than 100,000 miles (160,000 kilometres). It is here, at the base of the convective zone, that the Sun's magnetic field, which drives so much of the activity we saw higher up), is believed to be generated. The temperature outside is approximately two million degrees Celsius, but this is still not hot enough to power the Sun, and so we must descend further, under our own steam this time.

The Sun contains roughly 99 per cent of the Solar System's mass, and is 865,000 miles (1,392,000 kilometres) in diameter, and therefore large enough to contain more than a million bodies the size of the Earth. It has no solid centre, but is gaseous throughout; and the Sun's temperature increases with depth; by the time we reach the core it has risen to a massive 15 million degrees. It is here that solar energy is produced – but not in the way many people expect, because the Sun does not burn in the manner of a coal fire.

A Sun made of coal, sending out as much light and heat as the real Sun does, would be reduced to ashes in about a million years, but we know that the age of the Earth is about 4600 million years, and the Sun is certainly older than that. So, we must look for another energy

RIGHT: Illustration showing the magnetic field lines and the interior zones of the Sun.

RIGHT: A composite of the Sun showing the range of scientific research. The interior is from the Solar and Heliospheric Observatory (SOHO): blue/aquamarine represents rivers of plasma under the surface, which is itself imaged by the Extreme Ultraviolet Imaging Telescope. Both views are superimposed on a Large Angle Spectroscopic Coronagraph image, to block the centre of the Sun so the corona can be seen in visible light.

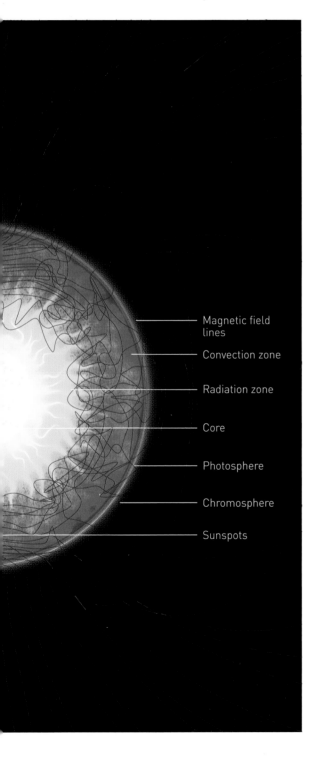

— Magnetic field lines

— Convection zone

— Radiation zone

— Core

— Photosphere

— Chromosphere

— Sunspots

source. The answer was found less than a century ago. The most abundant element in the universe is hydrogen, and, in fact, hydrogen atoms outnumber the atoms of all the other elements put together.

The Sun contains a great deal of hydrogen, and at the core, where temperatures and pressures are so colossal, remarkable things are happening – the nuclei of hydrogen atoms are merging to form atoms of the second lightest gas, helium. It takes four hydrogen nuclei to make one nucleus of helium, but in the process a little mass is lost and some energy released. It is this energy that makes the Sun shine, and the loss of mass, or 'weight', if you like, amounts to four million tonnes per second. This sounds staggering, and the supply of available hydrogen fuel will not last forever, but there is no immediate cause for alarm, because the Sun will not change dramatically for at least a billion years.

It is down here in the depths of the Sun that the energy on which life on Earth depends is generated; but it does not have an easy escape. The same density that allows nuclear reactions to take place also gets in the way of light escaping. A typical photon will bounce around the interior of the Sun for ten thousand years, hitting electron after electron before finally escaping to speed out into space.

We can escape a little more quickly. As Ptolemy travels back upwards through the Sun, we are accompanied by small particles called neutrinos, also taking the direct way out. Neutrinos are, as their name suggests, tiny neutral particles produced by the atomic reactions that drive the Sun. Neutrinos do react with matter, but only when they score a direct hit on the nucleus, preferring otherwise to pass through undisturbed. 60 billion Neutrinos pass through your thumbnail every single second, for example. With such a low probability of interaction, they, like Ptolemy, can escape directly from the Sun's nucleus, providing scientists back on Earth with more information about the Sun's core than is otherwise available.

FOLLOWING PAGE: The false-colour teal is good for showing sunspots such as this flare at top right seen on 13 March 2012.

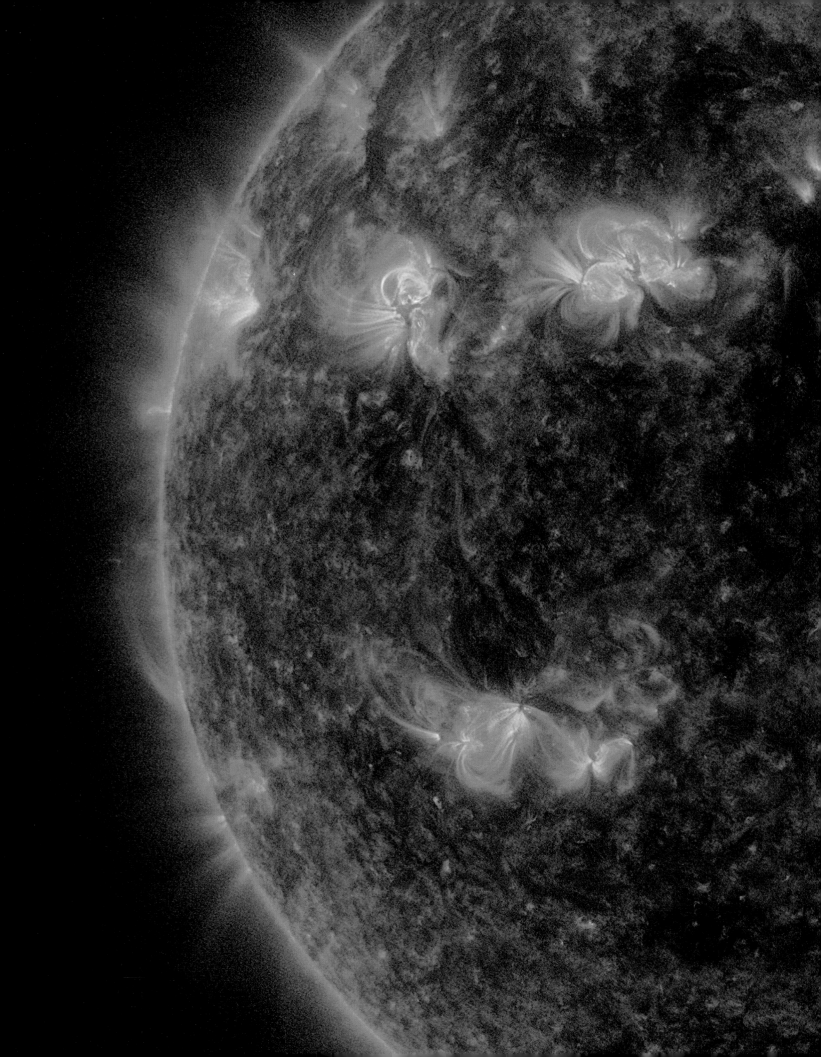

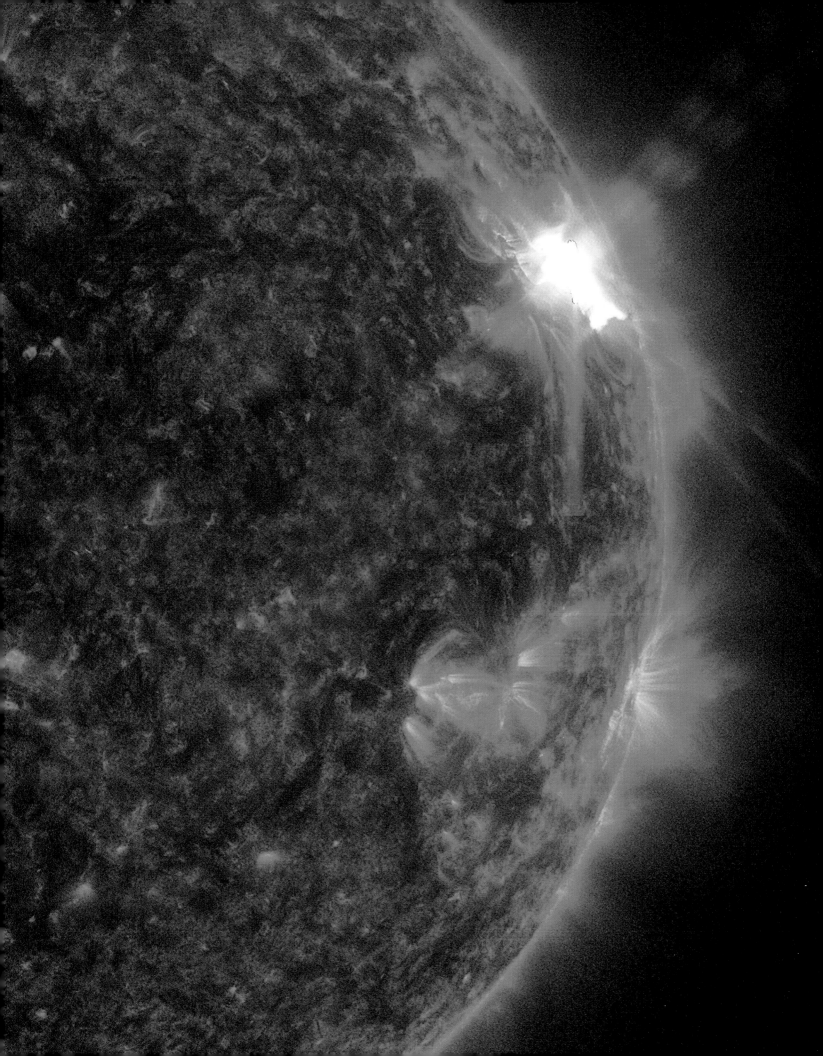

ENCOUNTER WITH A COMET

Distance from Earth: 8.3 light-minutes

The temperature drops steadily as we draw away from the Sun's centre. Suddenly, as we rise above the photosphere once more, the view from the windows clears. Outside the plasma of the Sun's interior, with its light-impeding electrons, light can reach us from great distances, and the rest of the Universe becomes visible once more. Ptolemy has taken us beyond the inner part of the Sun's atmosphere, and we are on our way to the first of the planets.

But look! Something very strange and beautiful has come into view. It looks like a luminous cloud with long, twin tails streaming out of it, and it's moving toward the Sun. We know this is a comet, and that its days are numbered; the cometary tails signal its eventual demise.

Comets are quite unlike planets. A comet has an icy nucleus – its only substantial part. Most of the comets that we see are moving around the Sun in very eccentric orbits, spending most of their time out in the frozen wastes of the Solar System, on paths which may take them almost out of our planetary system altogether. But when the comets sweep in closer to the Sun they are in danger.

The nucleus of a comet is something between a dirty snowball and an icy dirtball – a mixture of ice and rock that begins to come apart when exposed to the warmth of sunlight. As the comet is heated by the Sun, its material begins to sublimate – that is, turn from a solid directly into a gas, and is expelled as a tail.

Taking a close look at the comet alongside us, we can see that material is being expelled from a series of vents on the surface of the nucleus, throwing up a spray of material that forms the 'coma' – the luminous cloud of gas and dust that we could see from a distance. This material is swept away by the constant stream of charged particles that are expelled from the Sun, known as the 'solar wind'; and so the tail of a comet will always point away from the Sun. Actually, we should say 'tails' of a comet, because the gas and the dust behave differently under the influence of the pressure of sunlight and collisions with the solar wind, so most comets exhibit a white, curved, dust tail, and a relatively straight, blueish, gas tail.

ABOVE: A Kreutz (the name of the discoverer of this class of comets) sungrazer comet with a magnificent tail, plunges towards the surface of the Sun.

BELOW: Comet Lovejoy, a bright sungrazer, rather unexpectedly survived this 2011/12 close encounter with the Sun.

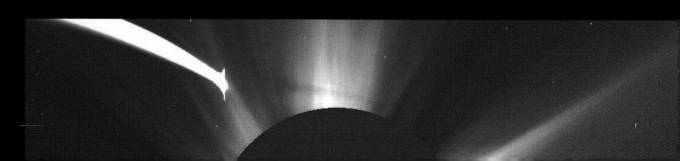

With such a dramatic loss of material, an average comet would not last long in the inner solar system before evaporating. Luckily, their eccentric orbits whisk them rapidly through the warm inner regions of the Solar System, and most comets survive to return again and again, as long as they stay on the path prescribed by their orbit. Many comets have perfectly stable orbits, and can last for a long time, but others, either on their first visit to the inner Solar System or disrupted by venturing too close to one of the planets, normally Jupiter, are in trouble. These sungrazing comets venture so close to the Sun that they are evaporated. The comet we are observing now is one of these; it is not going to escape, and, as it falls in toward the Sun's photosphere, it will be destroyed. We are seeing the last stages of the comet's existence.

Astronomers used to think that this sort of kamikaze dive was a relatively rare event, but satellites such as SOHO (the Solar and Heliospheric Observatory), which continuously monitors solar activity, have proved to have a pretty good second career as comet discoverers. Most comets seen in SOHO images are detected only on their final approach to the Sun, having been too faint or too distant from Earth to be spotted on the way in. Even so, the frequency with which comets enter the inner Solar System is much reduced from the days of the Great Bombardment that left the surface of the Moon so badly scarred.

What else can we find in these regions? Not a great deal, because the nearest of the planets to the Sun – Mercury – is 36 million miles (58 million kilometres) out. But, there is another possibility; in these torrid regions there may be small bodies, which are very difficult to detect from Earth, because they are always hidden in the glare.

Once, it was also thought that there might be a fairly large planet there, and it was given a name – Vulcan. A famous French philosopher, Urbain Le Verrier, even calculated its position in the sky, based on anomalies in the behaviour of Mercury in its orbit, which were later explained by relativity. Careful, but unsuccessful, searches were made for Vulcan but there are really only two ways of detecting it. During a total eclipse of the Sun, the sky becomes dark, and if Vulcan actually does exist it might be more discernible at this time. Alternatively, there could be occasions when Vulcan might pass between the Earth and the Sun, and would be seen as a black spot against the brilliant solar disc. Both Mercury and Venus (the only known planets which are closer than the Earth is to the Sun) do transit quite regularly, but Vulcan has never been observed in this way, and by now we are fairly sure that it is non-existent. Vulcan is one of the Solar System's ghosts.

During our trip outwards from the Sun we live in hope of glimpsing small bodies – a handful of asteroids do venture within Mercury's orbit – but they are few and far between and we do not succeed. These regions seem to be almost unpopulated, and we now point Ptolemy in the direction of Mercury.

LEFT: Comet Lovejoy is visible near Earth's horizon in this night-time photograph by NASA astronaut Dan Burbank, onboard the International Space Station (ISS) in 2011.

MESSENGER OF THE GODS

Distance from Earth: 8.64 light-minutes

Now we can start to see Mercury in the distance. Relieved to find the first solid ground since we left the Moon, we make plans to land there, but first it seems worth taking a look from long range. Astronomers on Earth did not know much about Mercury before the Space Age, because it is not easy to observe. Mercury always stays close to the Sun, and from Earth it is visible with the naked eye only when it appears low in the west just after sunset, or low in the east just before sunrise, so most people have never seen it at all. Mercury moves quickly across the sky from one night to the next, which is why the ancient Greeks named it after the fleet-footed scurrying messenger of the Gods.

Peering out through Ptolemy's window, we can see that Mercury really does look a lot like our Moon. There are mountains, craters and valleys, and the surface shows the same ancient, beaten-up look that is so familiar on the lunar surface. The diameter of Mercury is just over 3000 miles (4800 kilometres), and so, despite its status as an independent planet, it is not a great deal larger than the Moon.

Mercury's environment is different, of course. The mean distance of Mercury from the Sun is 36 million miles (58 million kilometres), compared with about 93 million miles (150 million kilometres) for our planet, Earth. But Mercury's orbit is a much more eccentric ellipse than that of the Earth, so its distance from the Sun actually ranges between 43 million miles (69 million kilometres) and only 21 million miles (34 million kilometres). This proximity produces surface temperatures which are extreme – very hot indeed during the day, and, because there is virtually no envelope of atmosphere, bitterly cold at night. At midday, on the hottest part of the surface, the temperature is so high that if you put a tin kettle down on the rocks the kettle would promptly melt. Mercury's 'year' – the time it takes to go completely

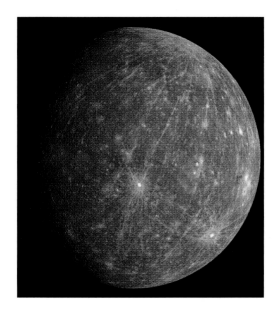

ABOVE: The 2008 MESSENGER mission to Mercury took this image 90 minutes after closest approach. The mission revealed the 30 per cent of the planet that had not previously been seen.

LEFT: Below and right of centre is a small crater with a pronounced set of bright rays extending across the planet's surface. Rays are commonly made in a crater-forming explosion when an asteroid strikes the surface. Their prominence implies the small crater formed recently since the rays have not been eroded.

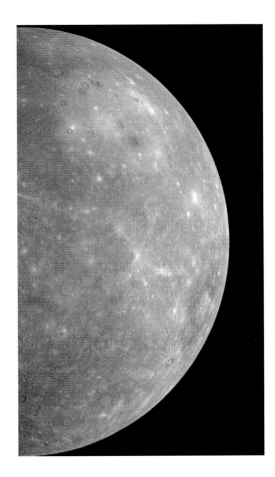

ABOVE: Colour differences on Mercury are subtle, but they reveal important information about the nature of the planet's surface material. A number of bright spots with a bluish tinge are visible in this image taken by MESSENGER in 2008.

around the Sun – is 88 Earth days, but the time it takes to make one complete spin on its axis is 58.5 Earth days. This leads to a very peculiar calendar, because, by the time the planet has made one revolution on its axis, the Sun has apparently moved! Days and nights on Mercury are thus very long; if you lived on Mercury the interval between one sunrise and the next would be 176 Earth days, or slightly over two Mercurian years! Unless they're planning a long stay, visitors are advised to head for the illuminated side, while the other remains in darkness.

The good news for travel agents is that there's no good or bad season to visit Mercury – the planet doesn't have seasons. Summer in the northern hemisphere of Earth happens in the months around June because the tilt of the Earth's axis points the North Pole toward the Sun during these months, whereas in December the axial tilt favours the south. Mercury's plane of spin isn't tilted with respect to its orbit, and so there is no seasonal variation.

It has always seemed clear that Mercury can have very little atmosphere, because its gravitational field is so weak. However, observations from the MESSENGER spacecraft, which is in orbit around the planet, revealed there is something there – sodium, potassium, calcium and other materials, all thrown up from the surface by collisions with solar wind particles or interplanetary dust. Its pressure is negligible, and if we scoop up a sample in a bottle, the bottle will be more 'empty' than that in an average laboratory's vacuum pump; but it is there, so it must be constantly replenished. Just as the solar wind forces material away from the comet we encountered on the way here, so it strips Mercury of its tenuous atmosphere. Mercury is a planet with a tail stretching away from the planet, much like a comet's tail.

There is a magnetic field here on Mercury – much weaker than that of Earth but quite unmistakable. Careful observation of Mercury's slow spinning supports the idea that this core is liquid, which was a surprise since calculations for such a small planet had led scientists to believe the core would have cooled and solidified. The challenge for scientists, rather than tourists like ourselves, is to explain how this small planet has maintained a molten core for over four billion years, since the formation of the Solar System.

We are close to Mercury now, and our best plan will be to ask Ptolemy to take us on a brief tour around the planet before landing.

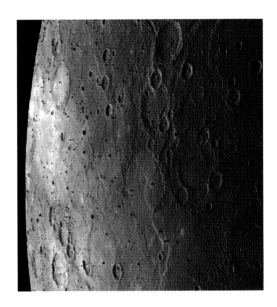

ABOVE: The limb of Mercury seen by the MESSENGER mission.

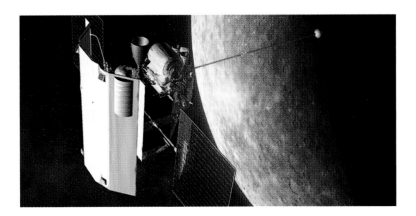

ABOVE: Artist's representation of MESSENGER at Mercury.

THE BASIN OF HEAT

Distance from Earth: 8.64 light-minutes

A good starting point for a tour of Mercury is one of the well-marked lava basins known as Caloris. About 1000 miles (1500 kilometres) across, Caloris has an interesting history, and it spent more than thirty years only half-discovered! When the first spacecraft to visit Mercury – Mariner 10 – flew past in 1974 and 1975 it was dawn or dusk at Caloris and the basin was half-hidden in darkness. MESSENGER revealed the rest of the basin, including one particularly prominent feature – a 25-mile (40-kilometre) crater close to the middle of the floor. From this crater a pattern of troughs radiates. The whole system looks so uncannily like a spider's web that the crater itself was called 'Spider' until the International Astronomical Union decreed that this was really too undignified, and christened it Apollodorus after a second-century Greek architect instead, while the troughs were designated Parthenon Fossae.

The Caloris Basin itself is, of course, an impact structure, formed around four billion years ago, from what must have been a colossal explosion, followed by extensive volcanic activity. The shock would have been felt over a very wide area. Caloris is surrounded by hilly terrain, with valleys and secondary craters. If we land there at noon, when the Sun is overhead, we will feel decidedly warm, because the thermometer will register around 430 degrees Celsius; we have chosen to explore the most torrid region of Mercury; indeed Caloris is Latin for 'heat'.

Mercury's orbit is markedly eccentric, and, like all the planets, it moves fastest at perihelion (the point where it is closest to the Sun), and much slower when far from the Sun. The speed at which the globe rotates on its axis does not vary, and there are thus two 'hot poles' where the Sun is overhead at noon. One of these hot poles lies in the Caloris Basin, and if we elect to stay there for 176 Earth days (twice the length of Mercury's orbital period, and three times its rotation period), we will be able to follow the sequence of events. The Sun will rise when the planet is furthest from the Sun, at aphelion, and it will ascend slowly, increasing its apparent size, until passing directly overhead; it will then stop, and move backwards for eight Earth days – because near perihelion the orbital angular velocity exceeds the spin velocity. The Sun will then resume its forward motion, shrinking in size until reaching the horizon, a quarter of a turn around the planet, setting 88 days after having risen.

If we had observed from a site 90 degrees latitude from Caloris, we would have seen the Sun rise at the time of perihelion, and there would have been no overhead hesitation, but, on reaching the horizon, the Sun would have set and then risen again, briefly, as if to say 'adieu'. After the second setting it wouldn't have been seen again for another 88 Earth days. One hates to think what a Mercurian would make of all this!

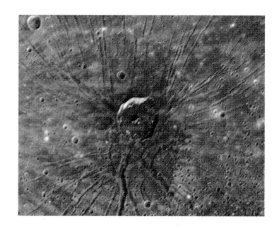

ABOVE: Skimming 124 miles (200 kilometres) above the surface of Mercury, MESSENGER revealed this never-before seen, spider-like feature in the depths of the Caloris Basin.

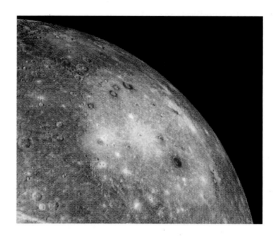

ABOVE: The MESSENGER spacecraft captured this colour image of the vast Caloris Basin region. Orange splotches around the basin's perimeter are thought to be volcanic vents, new evidence that Mercury's smooth plains are indeed lava flows. Other discoveries include evidence that Mercury has a global magnetic field generated by a dynamo process in its large core.

TOUCHDOWN ON MERCURY

Distance from Earth: 8.64 light-minutes

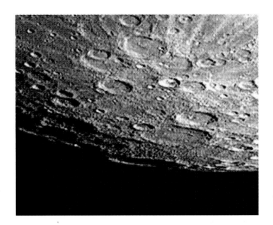

ABOVE: Mercury's South Pole was originally photographed by one of Mariner 10's cameras in 1974. The South Pole is located on the right-hand edge of the large crater that has only its rim sticking up into the light (Chao Meng-Fu Crater).

ABOVE: Chao Meng-Fu receives only grazing sunlight. MESSENGER revealed that much of Chao Meng-Fu's floor receives no sunlight at all. Some of the central peaks of Chao Meng-Fu, illuminated by the low Sun, can be seen peeking up from the crater's floor and casting long shadows.

Moving away from the blistering heat of Caloris, we are now heading towards a region of Mercury, just after dawn, where things are very different. Below us now is a crater known as Chao Meng-Fu, named after a thirteenth-century Chinese painter and calligrapher, though the landscape is a far cry from that of Imperial China. The crater is huge, around 100 miles (160 kilometres) across, and, like many other large Mercurian craters, it has a ring of mountains surrounding its centre rather than a single, central peak. We can see its very high walls shining brightly in the sunlight, but, because the Sun never rises high in the sky around here, the crater's interior is permanently dark and forbidding. In fact, roughly 40 per cent of the interior never receives any sunlight, rather like Cabaeus at the South Pole of Earth's Moon, and the floor of the crater appears very dark to us as we look down from Ptolemy's viewing ports. But Chao Meng-Fu has a special secret – it is a radar hot-spot!

When radar beams are used to probe Mercury's surface, instead of visible light, this otherwise obscure crater shows up as a very bright feature. However, we didn't need to come to Mercury to realize this; when astronomers back on Earth used giant radio telescopes, for example, the Arecibo telescope in Puerto Rico, which is 1000 feet (305 metres) in diameter, to bounce signals off the planet, they discovered radar-bright patches near both the northern and southern poles. However, this crater outshone them all, suggesting that deep within it lies an unusually reflective surface. What could the surface be?

The most popular theory is that it is ice that is lurking in the depths of Chao Meng-Fu, just as it does in our Moon's southern craters. The idea of ice on such a hot world as Mercury may seem unlikely, but the temperature at the bottom of this crater will never rise above –171 degrees Celsius, more than cold enough to keep a frozen lake in place. The mystery is not quite resolved; the reflections were not quite bright enough to indicate pure ice; so perhaps it is mixed in with the silicate rocks that make up most of the rest of Mercury's surface. But where could the ice have come from? It seems to be relatively recent, or at least ice that is more recent than the formation of Caloris, as younger craters shine more brightly in the radar maps. Perhaps the ice is produced locally, the result of chemical reactions initiated by the harsh sunlight in Mercury's regolith (the loose surface dust covering the underlying rocky core), and then somehow trapped in the dark crater. Alternatively, it may have been deposited directly in the crater by comets like the one we saw on the way here, plummeting sunward.

These questions may be answered by MESSENGER and by the Bepi-Columbo spacecraft that will soon scour the surface of Mercury. Trying to decipher the history of this fascinating world will keep scientists busy for a while. But we have other calls to make, and Ptolemy is already whisking us through interplanetary space, toward Venus.

THE SHROUDED PLANET

Distance from Earth: 9.45 light-minutes

Even from tens of millions of miles away, it's clear that Venus is a very different world from Mercury. Venus is a larger planet; in fact it is almost exactly the same size as Earth, and plainly it has an atmosphere. So, instead of the airless, barren and cratered landscapes of the Moon and Mercury, if Venus's atmosphere were transparent we might hope to see lands, seas, forests and deserts.

But all that's visible as we approach is a bright, virtually featureless disc – we are looking at the top of a thick layer of clouds. This sight baffled the first telescopic observers, who spent a lot of time chasing shadows in an attempt to pin down features on the planet's surface. In the absence of anything definite, speculation (sometimes scientific and sometimes less so) was rife. One feature in particular demanded an explanation: the 'Ashen Light' was a mysterious glow seen by some observers on the night side of Venus. We see a similar effect with the crescent Moon – often called 'the Old Moon in the Young Moon's arms' – but that is caused by reflection of light from the Earth, something that cannot apply here. A favourite (although inaccurate!) explanation was proposed in the 1830s by German astronomer Franz von Paula Gruithuisen, who claimed that the light appeared as the result of Venusian celebrations following the election of a new ruler.

Gruithuisen was at least correct in that the explanation for the Ashen Light lies on the hard surface of Venus, but, before we get there,

BELOW: Ultraviolet image of the clouds of Venus as seen by the Pioneer Venus Orbiter in 1979.

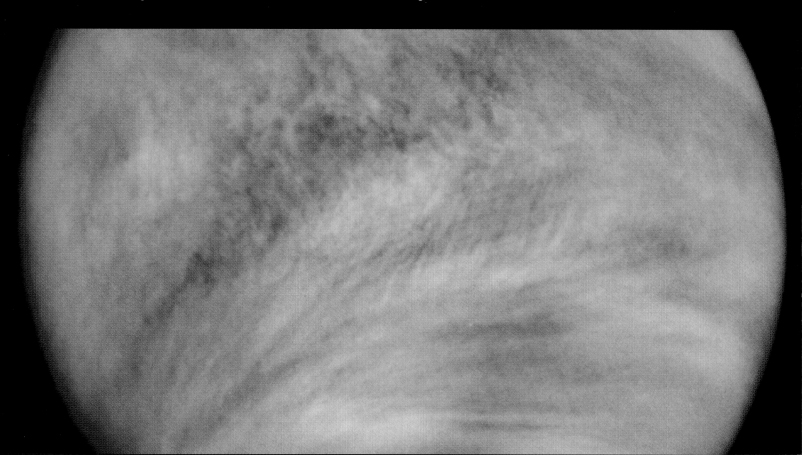

let us look at the composition of the atmosphere itself. The atmosphere is mostly carbon dioxide, with less than 4 per cent made up of nitrogen. The remaining lesser components of the Venusian atmosphere, including sulphur dioxide and even sulphuric acid, must make it a particularly unpleasant place.

Since the clouds of Venus block our view if we use normal visible light, it's helpful to switch to Ptolemy's infrared detectors, so that we can 'see' the heat emissions of the planet. Viewed in the infrared rays, the cloudy atmosphere glows gently, with a signature in its spectrum due to nitric oxide, first detected by the European Space Agency's Venus Express spacecraft. But, in this infrared light, the surface of the planet itself, heated to a temperature of over 450 degrees Celsius, is seen to glow too. And it is now believed that it is this glow that is responsible for the 'Ashen Light' mentioned earlier.

Moreover, in the infrared glow of Venus's surface, there are some areas of structure visible, with a few distinctive hotspots shining out, giving us an idea of the planet's geography.

Now that we can actually see the surface of the planet, one of the fundamental mysteries of Venus is apparent. The planet is spinning slowly, rotating on its axis once every 243 Earth-days. It takes 225 days to complete an orbit, so its 'day' is actually longer than its year. The interval between one sunrise and the next is 118 days. If that isn't strange enough, then consider the fact that Venus is rotating backwards compared with the spin of most of the other planets of the Solar System (Uranus also has a retrograde motion). How can this have happened? One possibility for the backward rotation is that the planet started off tilted on its side – perhaps as a result of collisions early in the process of planet formation. From there, chaotic effects and the drag of its heavy atmosphere may have done the rest. Although this theory is promising, we are far from sure about what actually happened.

We have used infrared light to reveal something of what is inside the shroud of Venus; but we have the spaceship Ptolemy – so let's now go in close, dive through the clouds, and take a proper look.

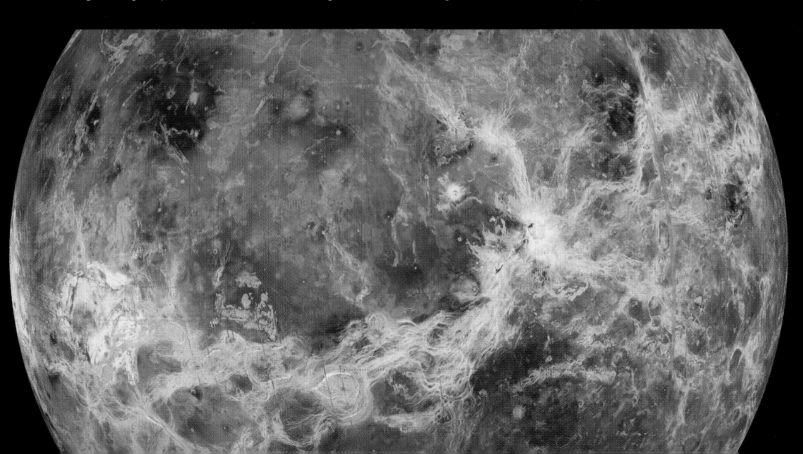

BELOW: To see below the clouds of Venus, in the early 1990s the Magellan spacecraft imaged the planet with radar and produced this high-resolution image of the surface. The bright area running across the middle is the highland region Aphrodite Terra.

BENEATH THE CLOUDS OF VENUS

Distance from Earth: 9.45 light-minutes

Descending through the Venusian atmosphere, it's the yellow clouds of sulphur dioxide that first catch our attention – not just because they produce sulphuric acid rain, making conditions on the surface distinctly unpleasant, but because they really shouldn't be there.

Sulphur compounds in the Earth's atmosphere are quickly removed by interaction with the surface rocks, so why doesn't the same process happen here? It may be that the composition of the surface of Venus is very different, making the absorption a slower process, but the alternative is that there is a source of sulphur dioxide somewhere on the planet. The likely candidate is a large and active system of volcanoes.

Further evidence for active volcanoes on the planet can be seen once we emerge beneath the clouds. Cruising over the surface of Venus we can see lava plains and mountains that look distinctly like volcanoes, topped by what must be enormous caldera (cauldrons left by volcanic activity). There must have been vulcanism here at one time. The distinct lack of impact craters (only about 900 have been identified across the entire surface), suggests that lava has flowed here relatively recently. Venus bears few of the kind of scars that cover the Moon and Mercury, and even if we take into account the corrosive

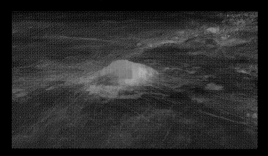

ABOVE: The volcanic peak Idunn Mons in the Imdr Regio area of Venus. The topographic backbone (brown colour) was derived from data obtained by NASA's Magellan spacecraft and the overlay was derived from data from ESA's Venus Express spacecraft.

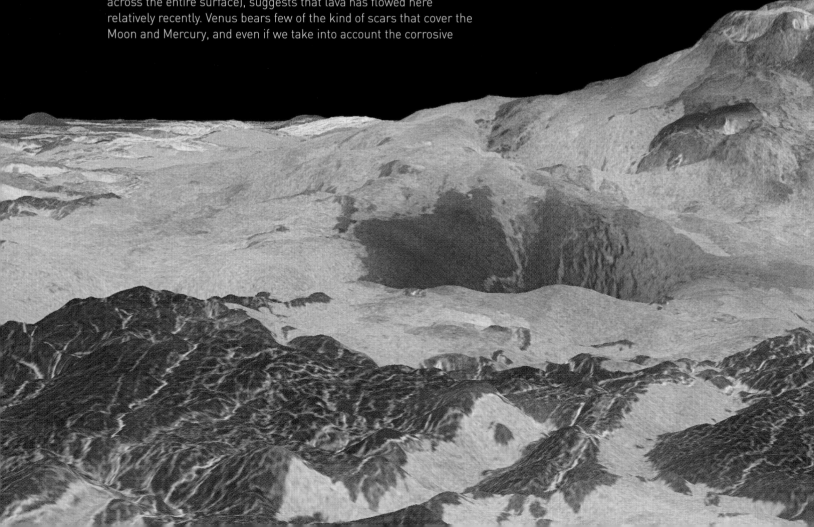

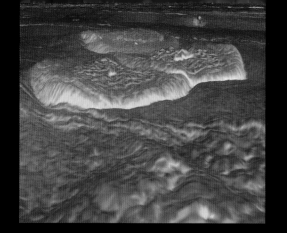

ABOVE: A portion of the eastern edge of Alpha Regio seen in this three-dimensional perspective radar view from the Magellan spacecraft. The three hills that are visible may be the result of thick eruptions of lava.

BELOW: The Venusian volcano, Maat Mons, mapped by Magellan using its onboard radar. Maat Mons is 5 miles (8 kilometres) high.

erosion by its acidic atmosphere, we are still left with the suspicion that Venus still has active volcanoes – but where are they? This is where those hotspots we saw in infrared light come in; we now aim our ship toward one of the brightest, Idunn Mons in the southern hemisphere, named after the Norse goddess of apples and youth. Idunn Mons certainly looks like a volcano as it rises majestically from the plain, surrounded by bright material which seems to flow away from the crater at its top. In fact, Idunn Mons looks almost exactly like the volcanoes of Hawaii's Big Island back on Earth, where lava has built an entire island topped with dramatic volcanoes. On Earth, movements of the tectonic plates typically shift the crust above the upwelling material, producing features like the Hawaiian island chain, as the hot lava repeatedly punches through the crust. Here, there are no plates to move, and so the volcanoes grow to be very large indeed.

Idunn Mons is different in composition from the neighbouring rocky surface, something that we can determine by carefully measuring variations in the gravitational pull of the planet as we fly over it. And the fact that we can see it glowing steadily in the infrared is a sign that the material is relatively recent – new enough for it not to have been weathered by Venus's atmosphere.

Studies conducted by Venus Express suggest that the most recent eruption could have occurred anything from hundreds of millions of years ago to just a few hundred years back. A recent eruption would suggest that another may occur soon. Is this the last gasp of a grand volcanic event that resurfaced the entire planet, or are volcanic eruptions just a normal part of being an Earth-sized planet? Despite the deeply significant differences between our Solar System's twin planets, Venus and Earth, the study of Venus will help us to understand what is 'normal' and what is 'unusual' in our own world.

A MAN-MADE ECLIPSE OF THE SUN

Distance from Earth: 10 light-minutes

We three intrepid astro-tourists woke up this morning with a new thought; that before we leave the vicinity of our own planet Earth entirely, we can use it to fulfil a new function. We can arrange for the Earth to block the Sun, and generate our own, private, total solar eclipse. A total eclipse of the Sun is an awesome spectacle down on Earth, because it's our only opportunity to look into the inner Solar System without scattered sunlight – daylight – obliterating our view. Out here, in space, we have no light-scattering atmosphere to contend with, but without an eclipse it's obviously impossible to direct our gaze anywhere near the blinding Sun.

We steer Ptolemy to a position just far enough away from the Earth for our planet to appear the same angular size as the Sun, and we position our ship in the middle of the Earth's shadow, where the Sun's bright disc is just covered. We are rewarded with a most incredible view. With the body of Earth blotting out the dazzling yellow photosphere, just as the Moon does in an earthly total solar eclipse, we see instead a brilliant red ring of fire – the chromosphere – the ionized (electrically-charged) gaseous 'skin' of the Sun, shining in hydrogen alpha emission, the characteristic red colour of hydrogen, which gives this layer its distinctive hot-pink colour. On the outer edge of this ring we see a few spectacular, similarly coloured, flame-like eruptions – the solar prominences – leaping into view, each of them large enough to swallow up a hundred Earths.

Of course, we, in Ptolemy, have actually travelled through these regions on our way to the centre of the Sun, but now we are seeing them with a new and unique perspective. Surrounding the chromosphere and prominences is another domain. The delicate tracery of the beautiful, pearly-white solar corona – the Sun's atmosphere – reaches out as far as 15 solar diameters, and has a complex flower-like shape which varies according to how active the Sun is at the time.

BACKGROUND: The coronal and zodiacal light captured by the Clementine spacecraft with Earth superimposed to simulate the man-made eclipse seen from Ptolemy.

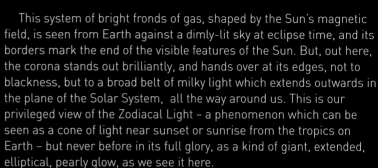

This system of bright fronds of gas, shaped by the Sun's magnetic field, is seen from Earth against a dimly-lit sky at eclipse time, and its borders mark the end of the visible features of the Sun. But, out here, the corona stands out brilliantly, and hands over at its edges, not to blackness, but to a broad belt of milky light which extends outwards in the plane of the Solar System, all the way around us. This is our privileged view of the Zodiacal Light – a phenomenon which can be seen as a cone of light near sunset or sunrise from the tropics on Earth – but never before in its full glory, as a kind of giant, extended, elliptical, pearly glow, as we see it here.

What is the Zodiacal Light? As long ago as 1683, French astronomer Gian Domenico Cassini suggested that its appearance was due to sunlight scattered by a thin, lens-shaped cloud of dust around the Sun, which was concentrated in the plane of the planets, and surrounding them. Following his theory the Earth was situated some way out toward the edge of the 'lens'. Much research and argument has gone on over the years, and our picture of the dust cloud has been refined, but current opinion is that Cassini got it largely right. However, the question of the exact composition and origin of the dust is still under debate.

It seems certain that a proportion of the dust originated in collisions in the asteroid belt, with the debris slowly spiralling in toward the Sun over tens of thousands of years. A further source is material shed by comets; as they make their periodic approaches to the Sun, dust particles and ices are 'baked' off their surfaces to form the spectacular cometary gas and dust 'tails' we see as the comets pass by. What has usually been missed is that there must be some component of dust that is simply passing through.

Our Solar System is travelling at about 19 miles (30 kilometres) per second through local space, toward a spot in the sky near the bright star Vega (Alpha Lyrae) – known as the Solar Apex. So, if we see dust moving at this kind of speed in a direction opposite to this, we might certainly begin to believe we are seeing interstellar dust particles. The proportion of this material may well be variable in time, depending on the nature of the interstellar medium we are passing through, but also depending on the ability of the Sun's magnetopause to block the particles, which are presumed to be electrically charged.

One of us (BM) is getting quite excited at collecting particles of dust out here, because he thought he detected an interstellar component in the Zodiacal Cloud in the 1970s, but neglected to finish writing up his thesis, at least until the twenty-first century.

ABOVE LEFT: The Zodiacal Light photographed at the European Southern Observatory (ESO) at La Silla, Chile.

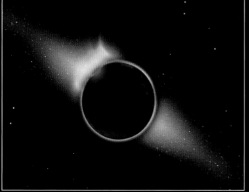

LEFT: Ptolemy's man-made eclipse reveals the dramatic extent of the Zodiacal Light in a way that is not possible during a solar eclipse seen from Earth, where the view is restricted by reflected light.

A GRAIN OF DUST

Distance from Earth: 10 light-minutes

The light scattered around by the Zodiacal Dust cloud is interesting to many astronomers only as something that sometimes gets in the way of their attempts to see into deep space. It is necessary to be able to subtract this 'light pollution' from their observations, so this gives a good reason to know its shape accurately. The actual study of the dust for its own sake was neglected for many years, as if dust were, as on Earth, nothing more than a nuisance.

But recent studies of an ever-increasing number of exoplanetary systems have revealed that dust clouds are an integral part of any evolving star and its family of planets. Indeed, these humble dust grains are the very building materials from which our bodies, and everything around us, are made. Space is never empty. Dust is everywhere, being created and destroyed, as part of every birth and death in the Universe: 'dust to dust'.

Can we collect some? Yes, Ptolemy is equipped with excellent dust-collecting capabilities, and we can put some grains under the microscope. The grains are very varied – some are made of material similar to that found in the rocky meteorites we see in museums on Earth, whilst others are more metallic. But they almost all have a similar structure; they are loose aggregates, much like this specimen which was actually picked up near the tail of a comet. Spongy in appearance, the grains are often called 'fluffy' particles. This structure, it is now thought, is a crucial part of the grains' role in being a catalyst in star formation. Their irregular surfaces allow chemicals to combine as they attach themselves, and when radiation blasts the particle, the necessary bonds are made to create the stuff of which stars are formed.

So although the dust particle is not the grandest spectacle we have encountered on our cosmic trip, our acquaintance with it is a very significant moment in our journey.

TOP: Space dust gathered in the stratosphere at an altitude of 6–30 miles (10–50 kilometres). The particle is about 10 micrometres across.

MIDDLE: Comet Lovejoy shedding dust from its tail. One of the Very Large Telescope's four domes is seen in the foreground.

RIGHT: Halley's Comet passes in front of the Milky Way in 1986. Unusually, it is named not for its discoverer but for the man who predicted its return.

A NEAR MISS?

Distance from Earth: 10.4 light-minutes

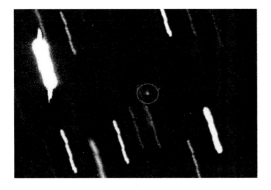

ABOVE: Asteroid Apophis, circled, en route to a close encounter with planet Earth.

ABOVE: Trees were knocked down and burned over hundreds of square kilometres by the Tunguska meteoroid impact.

ABOVE: Close-up of the asteroid 25143 Itokawa as seen by the Japanese robot probe Hyabusa.

Following our examination of Zodiacal Dust, we can continue to head out, away from the Sun. As we pass through Earth's orbit, we notice a few wandering bodies that warrant a visit. These are the near-Earth asteroids, and one of them may, one day, grab everyone's attention in the most dramatic way possible – by blazing through our atmosphere and impacting upon the Earth's surface.

Fortunately for us, such impacts by large bodies on the Earth are relatively rare events. While meteorites do hit us from time to time, it has been more than a century since the last major impact, which, luckily, took place in a remote and uninhabited region of Siberia. This event in Tunguska was powerful enough to fell trees over an area of more than 800 square miles (2000 square kilometres), giving some idea of the devastation that's likely to occur if a similar wandering rock should hit a major city.

Let's visit one – asteroid Apophis. Roughly 984 feet (300 metres) across, it's an unprepossessing sight, with a rocky, irregular surface pockmarked by craters, caused by the collisions it has endured in its life, out here between the planets. Apophis isn't a solid lump; it's more like a rubble pile; maybe as much as 40 per cent of its interior is made up of empty space, which gives it a strange, pebbly appearance, reminiscent of a day at the beach. Samples of material from Apophis would reveal chondrules – millimetre-sized blobs of mineral that must have once coalesced in free space – and such a body is called a chondrite.

These chondrites are, in fact, the most common form of asteroid, and Apophis is very similar to asteroid Itokawa, which was recently targeted by the Japanese 'sample return' mission Hayabusa. Apophis may one day soon receive a visit (and it might be less friendly than mere tourism), as it is being considered as a target for the European Don Quixote mission, which will test the possibility of deflecting hazardous asteroids by impacting them.

Why all this interest in Apophis? The reason is that it poses a possible threat – many of us will experience a very close encounter with it in the near future. Apophis will pass much closer to us than the Moon, within the orbits of geostationary satellites, only 22,236 miles (35,785 kilometres) above the surface of the Earth. Ominously enough, the pass is due on Friday 13th April 2029, but we know the asteroid's orbit well enough to rule out the possibility of a collision. This near miss will allow Earth's gravity to alter the asteroid's orbit, making it slightly harder for us to predict subsequent passes. It will come close to the Earth again in 2036, and at present, it seems there is something like a 1 in 45,000 chance that, on this particular occasion, it will hit us. Good enough odds, certainly, for us to sleep soundly in our beds.

For now, we will leave Apophis behind and look forward to a close view when it streaks across Earth's skies in 2029.

STOP-OFF AT DEIMOS

Distance from Earth: 14.1 light-minutes

The next planet in our sights is Mars, and we will need a suitable vantage point from which to survey this most interesting of planets. Mars boasts two small moons, named Phobos and Deimos; (after two junior Greek gods of panic and fear, sons of the God of War), and these will be perfect places to view it from, so we will head for one of them. If and when astronauts visit Mars, they will probably do the same, and various mission concepts along these lines are now being seriously studied by the world's space agencies. If nothing else, these moons would be amazing places to sit and operate robots down on Mars' surface, free from the delays in communication endured by Earthbound rover drivers today.

As we approach Deimos, the smaller of the two Martian moons, we marvel at its strange shape – which is rather like a distorted pear. This little moon is only nine miles (15 kilometres) across at its widest, and so does not possess enough gravity to mould itself into a sphere. Deimos orbits 12,500 miles (20,100 kilometres) from the Martian surface, only a tenth of the distance of our Moon from the Earth; it therefore moves very quickly, completing a circuit of the planet in just thirty hours. We can see a few large craters, two of which are named Swift and Voltaire, honouring two eighteenth-century authors who, oddly, both suggested Mars should have moons long before any were

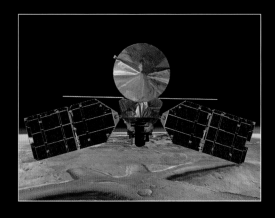

ABOVE: The Mars Reconnaissance Orbiter (MRO) above the Martian landscape.

BELOW: These colour-enhanced views of Deimos, the smaller of the two moons of Mars, were taken by the High Resolution Imaging Science Experiment (HiRISE).

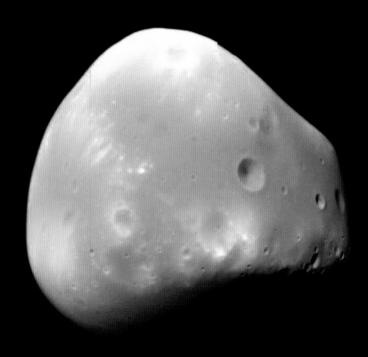

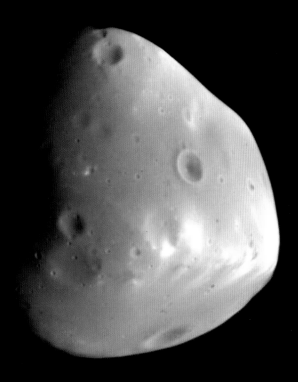

ABOVE: Mars Reconnaissance Orbiter's colour image of Deimos's sister moon, Phobos.

BELOW: Stickney Crater on Phobos, imaged by Mars Reconnaissance Orbiter.

actually observed. However, overall, the little moon looks strangely smooth; and this is because the whole surface is covered with a layer of loose regolith, a fine dust that coats both Deimos and Phobos.

A quick scan shows that the composition of Deimos's surface is a good match for one of the more populous families of asteroids, and indeed the theory is that it is nothing more than an asteroid that strayed too close to Mars and was captured by the planet's gravity. If that were the case, we might expect to find Deimos in a highly eccentric, elliptical orbit, but actually its path around Mars is almost a perfect circle. Phobos, closer in and also thought to be a captured asteroid, is in a similar orbit, but since it is closer to Mars tidal forces would have produced a circular orbit regardless of the starting conditions. Deimos's orbit remains a mystery.

We get a good look at Phobos as it whips round on its inside track, orbiting Mars just once every seven and a half hours. It is a strange world, with a chunk missing where a large crater, Stickney, records a past impact that must have almost shattered the fragile moon. Although the composition of Phobos matches that of a normal asteroid, like Deimos its density seems much too low – could it be hollow? Is Phobos just a loose, dusty pile of rubble masquerading as a solid object, or might it harbour a reservoir of ice, a suitably low-density substance?

We don't know, but enough doubt remains to provide support for an alternative theory for the formation of Phobos and Deimos. Perhaps they aren't captured asteroids at all, but were formed in situ from material either left over from Mars' formation or thrown up by subsequent impacts. This is certainly possible – think of the impact that produced the Earth's Moon – but it fails to explain exactly why the composition of Mars' two attendants is quite so asteroid-like.

As for us, we shall leave the study of Deimos and Phobos, and turn our attention to the Red Planet itself.

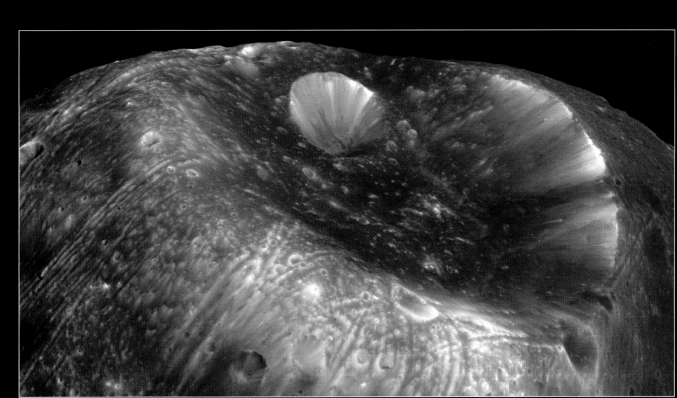

THE WAY TO MARS

Distance from Earth: 14.1 light-minutes

From Deimos, Mars is an awesome sight in the sky – covering an area which is a thousand times larger than the full Moon in an Earthly sky, and appearing one hundred times brighter than the full Moon appears to us from Earth. Mars is named after the Roman god of war; on Earth its blood-red colour in the night sky was noticed centuries ago. But from here, only a few thousand miles from its surface, we can see that the colour isn't uniform. The wide expanses of reddish-ochre are dotted with dark patches here and there, while the polar regions are covered in brilliant white caps.

The polar caps look like ice, and indeed they are, as we shall see. The northern one is rather larger than the southern - a consequence of the seasons, for when it is summer in the south of Mars it is winter in the north, and the polar caps shrink and grow over the course of the Martian year. The reason Mars has seasons is that its axis is tilted at 24 degrees – not very different from the familiar 23.5 degree tilt of our own Earth – and the length of the day is similar, too, at 24.5 hours. It will be interesting to see how future colonists of Mars adapt to this longer day; one excellent suggestion (from the science-fiction writer Kim Stanley Robinson) is to keep the 24-hour clock, but have a 'time slip' – half an hour a day in which it is forbidden to schedule anything!

At times dust storms that begin down in one of the planet's valleys expand to obscure the entire surface. This is frustrating for astronomers, but downright lethal for solar-powered rovers down on the surface which rely on a clear atmosphere to receive enough light to continue moving. Right now though, as we survey the Martian surface, while a few local storms are visible, things are generally clear.

Mars' tourist attractions are all visible. It is a world of staggering variety with craters, mountains, huge volcanoes and enormous valleys. A clear difference exists between the smooth northern plains, flattened by lava that must once have flowed across this surface, and the southern highlands, covered in craters and obviously an older surface. Quite why there should be such a dramatic difference is still not understood, but evidence is mounting that there was once a massive impact by something large – maybe half the size of Earth's moon. It seems that wherever we go we find evidence that the early Solar System was a violent place.

Prior to that collision Mars must have been a very different place. From our vantage point, we can see a myriad signs that suggest that Mars must once have had liquid running across its surface. Riverine channels snake across the surface, expiring in features that look for all the world like classic river deltas. Features of all sizes show the

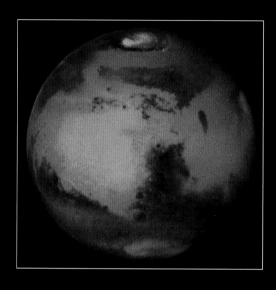

ABOVE: Mars seen through the eye of the Hubble Space Telescope.

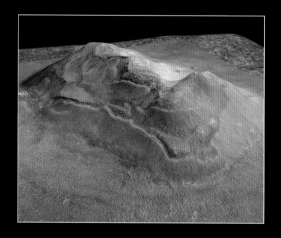

ABOVE: Mars Express's image of the Cydonia region, site of the famous 'face of Mars'. Sadly, this view shows it is just a hill.

ABOVE: Mosaic of the Schiaparelli (northern) hemisphere of Mars, a view similar to that which one would see from a spacecraft.

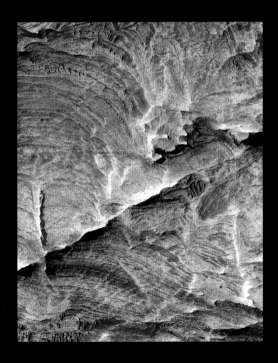

ABOVE: These white ridges were recorded on Mars. One theory is that they formed as water flowed through underground cracks and bleached and hardened the edges of surrounding rocks.

results of erosion by liquid, and we know there is water here, frozen in the Martian polar caps, which are made of water ice under a carbon dioxide (dry ice) crust. To have supported liquid water on the surface, the atmosphere must once have been thicker than that of Mars today, a thin layer predominantly composed of carbon dioxide. At its thickest, this atmosphere, still gradually escaping from Mars, sustains a pressure only one-tenth of that of Earth's atmosphere at sea level.

For more evidence of this formerly warmer, wetter Mars we will have to descend to the surface. Before we do so, we can use radar to study the ice caps, looking for evidence of climate change in the ice laid down year upon year and millennium upon millennium, just as Earth's scientists do by extracting cores of ice from the Antarctic. Other spacecraft have tried this experiment before, and both the European Mars Express and American Mars Reconnaissance Orbiter carried radar equipment. What we see is not a smooth distribution of ice, but layers of ice separated by other material. This seems to suggest that the story of Mars' climate is not one of slow decline from a warm, wet past, but rather one of cyclic change.

In other words, sometimes Mars is cool enough to deposit ice at its poles, as it is now, but during other periods it may be warm enough at the poles to lose ice, which would sublimate into the atmosphere, temporarily thickening it and perhaps allowing water to flow on the surface. One possible cause for such dramatic changes is that Mars may be unstable on its axis. The tilt we mentioned earlier may vary wildly over the course of millennia, causing melting at the poles when the tilt is large and they are exposed to the full glare of the summer Sun. There is no evidence for such dramatic swings of the Earth's axis, probably because we have the Moon to stabilize us; without it, our planet might also have swung as dramatically as Mars seems to do.

The study of Mars' climate is fascinating in itself, but it is also critical to the hopes of those who are examining the planet for signs of indigenous life. When calculating the chance that there might be life elsewhere in the Universe, the biggest unknown factor is what the odds are for life getting started. Our knowledge of life on Earth doesn't help, because without life already established here we wouldn't be around to ask the question. It's tempting to think that a wet Mars with a thick atmosphere might have most of the necessary ingredients, but then again, although a lot is known about how life evolves once it has begun, the conditions for the emergence of the first life-form are still not known. A negative result won't tell us much, but if some form of fossil, let alone a living organism, could be found somewhere on Mars, then the probability of life being scattered throughout the Cosmos increases dramatically.

In fact, one announcement of fossil life from Mars has already been made, by scientists studying a meteorite, recovered in Antarctica but which has been shown to come from Mars. A group of researchers claimed in the 1990s that tiny structures found deep within it resembled terrestrial bacteria, albeit one-tenth of the size. Most scientists are sceptical of this assertion, but debate rumbles on. One thing everyone agrees on, though, is that there is ample reason to explore Mars thoroughly. Time for us to land.

THE HOURGLASS SEA

Distance from Earth: 14.1 light-minutes

To start our exploration, let's descend to the feature most familiar to telescopic observers of Mars over the centuries – a large, dark patch, roughly triangular in shape which they named the 'Hourglass Sea'. Not really hourglass-shaped, and actually not a sea, either, this is the Syrtis Major Planitia, a smooth, high plain, which is believed to be the remnants of an ancient shield volcano. At over 800 miles (1280 kilometres) across, it is one of the most conspicuous markings anywhere on the planet.

What sort of place are we landing on? Well, it's decidedly chilly. Although Syrtis Major straddles the equator, the temperature never rises higher than a few degrees above zero Celsius, falling to freezing point long before sunset. The dark colour is a result of the underlying dark, basaltic rocks and relatively thin layer of dust. The lower-lying regions are covered with the reddish material typical of the Martian surface, primarily oxides, and fairly described as 'rust'. The same dust makes the sky a pinkish-yellow colour, rather than the familiar blue of Earth, although sunset produces amazing purple and deep-blue tones across the alien landscape. Looking at the sky, it's immediately obvious that the Sun is appreciably smaller and dimmer than it appears from Earth, adding to the alien feel. It's also eerily quiet, since the thin Martian air is not good at carrying sound waves.

Although we are on the surface of a substantial planet, spacesuits are essential for wandering outside. The problem is that, under the

ABOVE: A portion of a trough in the Nili Fossae region of Mars is shown in enhanced colour from the HiRISE camera on the Mars Reconnaissance Orbiter. This region northeast of Syrtis Major is rich in minerals.

BELOW: Mars Global Surveyor's view of Syrtis Major, the large dark area. This image was taken during the southern summer on Mars.

ABOVE: Mars Express returned these remarkable views of the Syrtis Major region on Mars that show lava flows that flooded the older highland material, leaving behind buttes – isolated hills with steep sides that were too high to be affected.

BELOW: The bright area at upper left is known as Arabia, and the dark area to the right is Syrtis Major Planum.

low atmospheric pressure here, liquids, including our blood, reach boiling point at a much lower temperature. Without a spacesuit, one's blood would come to the boil after only a few moments, just as it would in the vacuum of space – with unpleasant results for everyone.

Dangers aside, there is much to explore. We have landed in the northern part of Syrtis Major, just below the highlands of the Nili Fossae region, an ideal place from which to study many of the features that led us to conjecture, from orbit, that water must once have flowed across the Martian surface. We can wander along a dry river valley, which, branching here and there, is cut deep into the lava plains. It is large, often as much as 875 yards (800 metres) across (roughly half the width of the southern Mississippi). The valley disappears when it enters a deep crater, presumably the result of a more recent impact, but emerges again on the far side. As the surrounding landscape drops to meet the valley, we begin to see raised patches of rock which are teardrop-shaped – these were once islands, areas of harder material eroded into this characteristic shape by the action of a flowing liquid.

Such a dramatic landscape must have been created by a dramatic event, and that's what the geology of the area is telling us. The valley must have been carved, not by a gentle meandering river over thousands of years, but by a sudden flash-flood, with hundreds of millions of gallons of water released in a short period of time. That sounds like a lot – indeed it is a lot of water – but similar 'mega floods' are believed to have carved out the unique landscape of the Scablands in the state of Washington, back on Earth.

Where did the water come from? Although we will probably never be sure what happened, it may have come from the ice caps or, if water ice had been trapped under the surface of Mars, heating by volcanoes might have released such dramatic sudden floods. Evidence gleaned from the rocks suggests that the canyon formed roughly at the time when the rest of Syrtis Major was a volcanically active region, which certainly fits this theory.

There is other evidence for Martian volcanism, and we can take a close look at the grandest Martian volcano of all – Olympus Mons.

OLYMPUS MONS

Distance from Earth: 14.1 light-minutes

The Syrtis Major feature could keep any scientist busy for a lifetime, but there's a whole planet to explore. We cannot see a very wide area from our landing point; the smaller size of Mars relative to the Earth means that its globe curves more sharply, making the horizon appear closer. It's back in the ship, therefore, for the hop over to Olympus Mons (Mount Olympus), the highest mountain in the Solar System, dwarfing Earth's Everest, towering to a monumental 88,000 feet (26,822 metres) above the adjacent terrain from a base that is almost 350 miles (563 kilometres) wide. In fact, Olympus Mons is so enormous that it pokes up through the atmosphere, and its caldera is practically out in space; one day, it may make an absolutely wonderful site for an astronomical observatory.

Olympus Mons is, in fact, just one of the four volcanoes that sit in the raised portion of Mars known as the Tharsis Rise; the others are Arsia Mons, Pagonis Mons and Alcaeus Mons, all lofty enough for the summits to protrude above the uppermost layer of an average dust storm. There are plenty of other volcanoes upon Mars, but none rival the monsters of the Tharsis Rise.

The caldera of Olympus Mons itself is magnificent; 53 miles (85 kilometres) across and nearly 2 miles (3.2 kilometres) deep, it is a complicated place with six overlapping craters and numerous pits. Flanked by steep cliffs, it would present a challenging climb – more so than the volcano itself, which is likely to be a disappointment to serious mountaineers. The slope of the volcano is very gentle, and, given enough patience and stamina, it should be possible to walk all the way to the top.

This gentle climb is typical of the most massive shield volcanoes, which are produced by liquid lava that was expelled from the vents. Given the volcano's massive size, the lava expulsion process must have taken a long time, and what we are looking at is probably the result of countless eruptions. When did this process end; indeed, did it end? We cannot be sure that Olympus Mons is extinct, and it may be merely dormant; the lack of craters on and around its massive form is at least suggestive of fresh lava flows within the last few million years.

We've focussed on the possibility of recent events on Mars, but we need to get a clear overall view of Martian history – for that, we need a close look at the Martian surface.

ABOVE: Olympus Mons from the Mars Global Surveyor.

RIGHT: Clouds hover over the volcano peaks of the central Tharsis region. Olympus Mons is dominant at upper left.

BELOW: Olympus Mons, the largest volcano in the Solar System, has a volume fifty times the largest volcano on Earth.

BOTTOM: Mars Global Surveyor's image of Olympus Mons.

THE FLIGHT OF THE PHOENIX

Distance from Earth: 14.1 light-minutes

The frozen north of Mars – the planet's equivalent of our Arctic – is a desolate place at the best of times. Vast areas are covered by a nearly featureless plain, broken only by a strangely hexagonal terrain that, from orbit, looks like an enormous Martian equivalent of Northern Ireland's Giant's Causeway. These strange features are the result of the continual freezing and refreezing of the surface, causing the 'soil' to crack. Each winter this entire region disappears under the encroaching ice cap, only to reappear as the ice recedes in spring.

Astronomers were attracted to this unusual region as they had already detected what appeared to be a vast reservoir of ice under the surface by using radar from orbit. The signature is very strong, but for real, solid proof, we want to get our hands on the ice, to be able to sample it, touch it and test it. We're not the first to visit, though, and by skimming over the surface we can find the ghostly landing site of the Phoenix spacecraft. When Phoenix first landed on May 25, 2008, it would have been easy to spot; it can be seen in images from orbit, with its bright blue solar panels glinting against the ochre surface. Now, the winters have dulled its surface and it is a sorry sight, but it did incredibly important work.

Phoenix was designed to study the history of water and the habitability potential of the ice-rich soil of the Martian Arctic. In other words, it was looking for the raw material for life on the Red Planet. With only a brief summer in which to explore, there was no point making the spacecraft a rover (any part of this vast plain looks much the same as any other), and there was plenty to do while the constant sunlight of the Arctic summer could recharge its batteries, as the spacecraft continued to send data until the Sun finally set over Mars.

Phoenix is not particularly large, measuring 18 feet (5.5 metres) in length once its solar panels were successfully unfurled, with an 11-foot (3.5-metre) deck for the scientific instruments and a height of seven feet (2.2 metres) measured to the top of the mast. Its expected lifetime was an all-too-brief 90 sols (a sol is a Martian day, 24 hours and 37 minutes long).

The landing area is flat, and appears today roughly as it did in those first pictures from Phoenix, with a surface covered in small pebbles and cut by the small troughs that mark the boundary between the

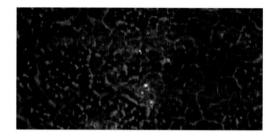

ABOVE: This 2012 image is of seasonal frost in the northern plains of Mars. As the frost sublimates (changes straight from ice to gas) hexagonal and polygonal patterns are revealed. The deceased Phoenix Lander can be seen in the centre.

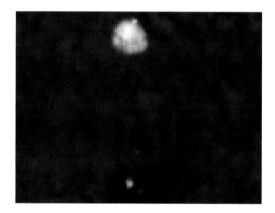

ABOVE: Phoenix was photographed parachuting toward the surface by the orbiting MRO, a remarkable technical feat from two craft millions of miles from home.

BELOW: Self-portrait: panoramic view of the Phoenix Lander on the Martian surface.

ABOVE: Phoenix Lander self-portrait.

SOL 20   SOL 24

ABOVE: Phoenix discovered ice when it dug down into the soil, as seen by the Surface Stereo Imager. In the four days between these shots some ice had clearly melted.

hexagons mentioned earlier. The spacecraft carried a robotic arm which could dig up to half a metre below the surface, scooping out material to be analyzed. It proved to be a clumsy device, struggling to get the slightly sticky, icy soil into its instruments, but, once it succeeded, a wealth of chemical information was obtained and sent back to Earth.

There was plenty of ice, which showed up first as a patch underneath the legs of the craft, presumably where the dust that covered it had been removed by the spacecraft's rockets. The 'soil' itself turned out to be mildly alkaline – around the right pH for growing asparagus, although the prospect of any Martian gardeners being around to take advantage of the opportunity any time soon seems remote! A real surprise was the presence of highly reactive molecules called perchlorates.

Phoenix was only able to sample the soil in one place, and arguments have raged since its discovery about how widespread these perchlorates might be, and what consequences, if any, they have for the odds on life getting started. On the one hand, the presence of reactive molecules might provide an energy source for the simplest lifeforms, whereas on the other hand we might expect the perchlorates to prevent the formation of life's complicated chemistry.

Whatever the truth, they may have solved one decades-old mystery. A previous generation of landers, the Vikings, had carried experiments which were uniquely, explicitly designed to look, not for the raw ingredients of life, but for the signs of life itself. The results were inconclusive, with samples of soil showing evidence of significant reactions when heated up, just as would be expected if the sample harboured microorganisms. Other experiments were negative, but all can now be explained if reactions with perchlorates were responsible for the first set of results.

Of course, there's no way to be sure that the composition of the soil in the Martian Arctic explored by Phoenix reflects that of the very different Viking sites. There was enough for Phoenix to do in its northern home, but the midnight Sun characteristic of midsummer could not last, and Phoenix saw it first dip behind a rock, and then the cold Martian nights drew in. As the nights grew longer, the temperature fell, and before long the Sun could not provide enough power to keep the spacecraft's solar batteries charged. The last signals arrived just over five months after landing, and the flight of the Phoenix was over. Cameras still watched from orbit, just as we do now, as the ice reclaimed its territory and then, once again, receded. The Phoenix is still here, its mission fulfilled, and it would be nice to think that one day it will be taken to a Martian museum. If not, as the winters pass all trace of Phoenix will disappear; its legacy – the scientific results it sent back to its home planet, Earth, and to the scientists that waited there. In the meantime, the site of the Phoenix remains an interesting spot for the curious visitor.

WE MEET SPIRIT AND OPPORTUNITY

Distance from Earth: 14.1 light-minutes

Craters are valuable to scientists because they allow us to look below the dusty Martian surface, reading the history of the planet in the rocks that make up a crater's sides and floor. We might try the crater Gusev, for example, a relatively undistinguished feature in the colder southern hemisphere of the planet. Just by flying over it we can see that it is most likely a lake bed, 99.4 miles (160 kilometres) across, fed by a long channel which cuts into the crater in the south. This watercourse, Mawrth Vales, is now bone dry, but once must have carried a raging torrent of water. There are numerous smaller craters scattered over Gusev's floor; these must be the result of impacts that have occurred in the few billion years since its formation, serving now as the signals that tell us that this is an old part of the surface. The floor is otherwise made up of volcanic rock, which has buried any sediment left from the crater's time as a lake.

Gusev, then, may look promising, but it doesn't offer the critical evidence needed to secure our view of Mars' watery history. Nonetheless we aren't the first to seek it here. A final pass over the surface reveals tyre tracks, beginning in the crater's centre and heading off towards a small group of hills, a mile or two away. Swooping close, we see a strange flower-shaped platform sitting in the crater's centre, surrounded by what looks like soft, silk parachute material. This is the landing site for NASA's Spirit rover, which touched down in 2004. Its designers sent it here because they, too, hoped for evidence of lake sediment at the floor of the crater, but while they were disappointed in that respect, Spirit delighted them, surviving for years beyond its planned 90-day mission.

We can follow Spirit's tracks, which have yet to be eroded by the thin Martian air, over towards the group of hills a few miles away. These are the Columbia Hills, named in honour of the crew of the tragically destroyed space shuttle. Spirit climbed to the summit of the 300-foot

ABOVE: A simulated image of one of the Mars Rovers on the surface of the Red Planet.

BELOW: Opportunity at the rim of Endeavour Crater.

ABOVE: The rocky outcrop, dubbed 'Longhorn' and behind it the sweeping plains of Gusev Crater. The Spirit Rover captured this view with its panoramic camera.

ABOVE: Spirit Rover's view of West Valley.

(91.4-metre) Husband Hill, to gain an excellent view of the terrain. It took 591 sols (Martian days), to reach the top. Leading down from the summit suddenly the tracks end, and there is Spirit itself. Still stuck in a sand trap that fatally ensnared it back in April 2009, Spirit sits silent, its mission now over.

However, at the speed of thought, let us quickly relocate to the other side of Mars, where we can visit Spirit's twin sister, Opportunity, which is still alive and well, roving onwards after more than seven Earth years on the Red Planet. It's at present exploring Meridiani Planum, using a series of craters, each larger than the last, to drill down into Mars' history. The cliffs exposed by the crater walls offer a history of Mars; as the rover moves further down, alongside each cliff, so we read earlier and earlier chapters of the planet's history.

Opportunity began its adventures on Mars in a crater, arriving in the Solar System space exploration equivalent of a golfing hole in one. The rovers parachuted down toward the surface, and then, were cut loose, surrounded by giant airbags. Bouncing across the surface, Opportunity came to rest within a small crater, but it soon moved on to bigger, deeper targets. Endurance, 426 feet (130 metres) across and 65 feet (20 metres) deep, was its first target, but by June 2007 the rover was poised on the edge of Victoria Crater, half a mile (0.8 kilometre) in diameter and 300 feet (91 metres) deep. After a preliminary tour of the crater's edge, Opportunity descended from a site known as Duck Bay. The descent was risky – standing on the edge of the crater now is enough to induce vertigo! – but it went well, and the rover's tracks lead us back out of the crater and toward an even larger crater.

This crater is called Endeavour, 13.6 miles (22 kilometres) across and 984 feet (300 metres) deep, with exposed rock that will tell us about millions, if not billions, of years of Martian history. The plucky little rover Opportunity, no bigger than a Mini, makes a strange sight as it struggles along; a broken wheel forces it to permanently drive backwards, and Opportunity's robot arm can no longer be stowed away, so it bounces around, held at an awkward angle. Opportunity is not fast, moving about 300 feet (100 metres) on a good day, but after a huge effort it is now exploring this geological jackpot.

Between them, the two rovers' geological investigations have established beyond doubt that much of Mars was once covered in water. The primordial oceans of Mars may not have been friendly places, though, as a lack of carbonates at the Spirit site in particular indicates the water may have been rather acidic.

BELOW: Mars Exploration Rover 'Opportunity' showing a view of Duck Bay in the western portion of Victoria Crater; the solar panels are prominent in the foreground.

LOOKING FOR WATER ON MARS

Distance from Earth: 14.1 light-minutes

Any good tour includes sites of topical interest, and Newton Crater might suddenly become an essential stop in any Martian tourist's itinerary – as the only place to see water flowing on the surface. In August 2011, scientists using NASA's spacecraft released information indicating possible evidence of salty liquid flows active on Mars today. The images, which are from the HiRISE (High Resolution Imaging Science Experiment) camera onboard NASA's Mars Reconnaissance Orbiter, come from several sites in the southern hemisphere, including the Newton Crater. The features are dark markings called, in the technical jargon, 'recurring slope lineae'; they are narrow – no more than five feet (1.5 metres) to 16 feet (5 metres) wide, and are seen on relatively steep slopes during warm seasons, with angles of 25–40 degrees.

Repeat imaging by HiRISE shows that the lineae markings or features appear and extend in length during the warm seasons and fade during the cold seasons. They extend down from the bedrock, are often associated with small channels, and hundreds of them form in rare locations. The lineae appear to lengthen during the southern spring and summer at latitudes between 48 and 32 degrees south, primarily on equator-facing slopes. Liquid brine near the surface might explain these features, but the exact source of the water is not understood. Salinity lowers the temperature at which water freezes, and water as salty as that in the Earth's oceans could exist in these sites, at least in the Martian summer.

It must be understood that these observations do not prove that salty water does actually flow upon Mars at the present time, but it does make it more likely, and this again increases the chances of finding some kind of life on the Martian surface. It has been suggested that there may be hardy microbes surviving in these short periods of summer meltwater, but we must be careful to avoid jumping to conclusions, and these observations, important though they are, are far from conclusive. Still, it is awe-inspiring to imagine standing at the base of one of these gullies, reaching down to touch what might be still-damp Martian soil.

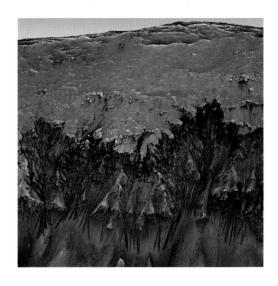

ABOVE: This image shows flows that appear in spring and summer on a slope inside Newton Crater.

ABOVE: The dark features (lineae) that extend down the slopes of Newton Crater can be explained by flows of fluids.

LEFT: Another view inside Newton Crater, with more evidence of flows – looking very much like a terrestrial river basin.

EROS

Distance from Earth: 13.9 light-minutes

ABOVE: Asteroid 433 Eros seen from the NEAR-Shoemaker probe.

ABOVE: 433 Eros in false colour. Here colour represents density, and red is material that is more dense.

The time has come to leave Mars and set a course for the outer Solar System, but this entails crossing the crowded asteroid belt that lies between the Red Planet and Jupiter. This is the home of hundreds of thousands of small bodies, ranging in size from dwarf planets, which are worlds in their own right, right down to rocks no larger than a pebble on the beach.

Most small bodies stay safely within what is known as the Main Belt, the part of the Solar System where these celestial vagabonds are most common; and soon after leaving the Martian orbit we're already encountering them. Among these celestial vagabonds is a familiar potato shape, the asteroid Eros, which has a particularly eccentric path that not only crosses inside the orbit of Mars, but also within 15 million miles (24 million kilometres) of Earth. Even if it is only 9 miles (14.4 kilometres) wide and 20 miles (32 kilometres) long, any collision with Earth would cause global devastation; but luckily there is no chance of that.

Approaching Eros, we can see numerous craters, pits, mounds and ridges. Dramatic mountains are too much to ask of this little asteroid; the smaller the body, the harder it is to support major features. The distance from Eros's highest point to its lowest is only about 1.2 miles (2 kilometres). Glinting within a saddle-shaped feature, known as Himeros (named after one of the winged love-gods that danced attendance on Eros in Greek mythology), lies the NEAR-Shoemaker spacecraft that was deliberately crashed here at the end of its mission in 2000. Like NEAR's constructors, we're struck by the remarkably uniform character of the asteroid's surface, with its lack of small craters.

Eros's eccentric orbit suggests that it may be a fragment of a previously larger body. A long ridge may mark a fracture, a break with whatever used to be attached to what is now Eros. The collision which caused the break-up may also have sent Eros spinning towards the inner Solar System, placing it in the chaotic orbit where we find it today. The impact of this collision may have had effects far beyond Eros itself; the composition of Eros is similar to many asteroids in the inner part of the Main Belt, suggesting a common origin. We shall see that the asteroids hold many more surprises, and as we pass through the belt, visiting some of the largest, we encounter next the bright world of Vesta.

THE BRIGHTEST ASTEROID

Distance from Earth: 20.6 light-minutes

Vesta is the brightest of all the asteroids, and so it's not surprising that it has been visited by a spacecraft from Earth. We will pause there in Ptolemy too. The Dawn spacecraft spent a year here, mapping the surface of this intriguing body which is, after all, big enough at 330 miles (530 kilometres) across to be as varied and interesting a world as any planet, rather than just another small asteroid. That isn't to say that Vesta is a hospitable place; its surface temperature ranges from about –20 Celsius to –100 degrees at the poles, and it is, of course, far too small to hold on to an atmosphere.

Vesta has a turbulent history. In the recent past, no more than a billion years ago, it seems to have lost one per cent of its mass in a collision. The event has left an enormous crater 290 miles (466 kilometres) in diameter, occupying much of its southern side, with a large protruding central peak, evidently made up of material that was excavated by the impact.

A lot of material must have escaped from this explosive event. Vesta is actually known to be the source of objects which have fallen to Earth, known as Howardite-Eucrite-Diogenite (HED) meteorites – so when we handle such a piece of rock, we may actually be handling a piece of Vesta. Can we identify other debris from the collision? Yes; several asteroids appear to have a suspiciously similar composition to Vesta; one of these is relatively large asteroid 192 Kollas, which probably has its origin deep inside Vesta's crust.

Rather than a true asteroid, it's perhaps best to think of this strange world as a failed planet. Like the so-called terrestrial planets, including Earth, Vesta is thought to consist of a metallic, nickel iron core and overlying rocky mantle in the surface crust. We can even make a reasonable guess that in its distant past, shortly after it had formed, it probably almost completely melted, allowing the metal core to separate from the rocky crust.

The source of heat is possibly a combination of aluminium-26 and iron-60, radioactive elements that may have been formed in nearby supernovae, those violently exploding stars that may have triggered the birth of the Solar System. These radioactive isotopes could have provided the extra heat which melted Vesta. Once they decayed, the asteroid would have cooled and solidified to its present state.

Vesta's huge southern polar crater is of special interest. Its width is 80 per cent of the diameter of Vesta itself, and its floor is about eight miles (13 kilometres) below the surrounding landscape, while the rim rises up to seven and a half miles (12 kilometres). As with Opportunity's exploration of Mars, a crater provides a chance to examine material from deep within Vesta. In this case, the crater has penetrated through several layers of crust, and possibly all the way down into the mantle. Our visit to Vesta has revealed a world unlike any other in the Solar System.

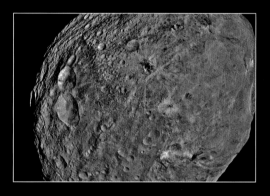

ABOVE: The heavily-cratered surface of Vesta seen from the Dawn probe.

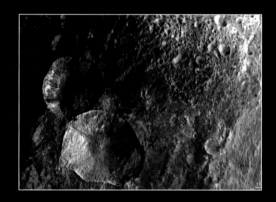

ABOVE: A set of three craters nicknamed 'Snowman', seen in the northern hemisphere of Vesta.

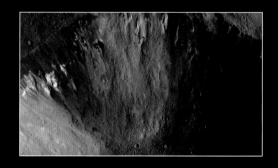

ABOVE: Dark and bright material in this crater

LARGEST OF THE ASTEROIDS

Distance from Earth: 23.8 light-minutes

ABOVE: Ceres as seen on 4 May 2015 in approximately true colour from the Dawn spacecraft. Two of Ceres' famous mysterious bright spots at Oxo crater and

Ceres is the giant of the Main Belt and is, indeed, now designated by the IAU (International Astronomical Union) as a Dwarf Planet; however, it is very much smaller than our Moon, with a diameter of no more than 600 miles (970 kilometres). Ceres orbits the Sun once every 4.6 years and shares a composition with the most common kind of asteroid, which accounts for roughly three-quarters of the Main Belt asteroids, and more than three-quarters of the asteroids here in the outer belt. Like most of its kind Ceres is dark, with a surface that's covered in craters.

No other Main Belt asteroid comes close to rivalling Ceres in size, and most are very small indeed. In fact, Ceres on its own accounts for more than a third of the mass of the entire asteroid belt.

We do not know a great deal about the makeup of Ceres; the best model we have is a rocky core overlaid by an icy mantle about 40–120 miles (60–200 kilometres) deep. This may well be composed of water ice and minerals such as clays. Obviously it is bitterly cold, and the temperature never rises as high as –40 degrees Celsius.

It is quite likely that Ceres once had an ocean below its crust, and there is even a slight chance that an ocean may still survive. In 2014 the Herschel Space Observatory observed several sources of water vapour emitted from the surface, and in 2015 the Dawn spacecraft

DEEP IMPACT

Distance from Earth: 32 light-minutes

Our next stop on our cosmic tour is at an ordinary little comet called Tempel 1 that orbits the Sun at a distance that varies from 12.48 light-minutes at perihelion (closest approach) to 39.1 light-minutes. If we look carefully, we may be able to spot a rare man-made crater on its surface, and this is what we've come to see. The fame of this comet is a result of it being visited by not one, but two spacecraft, one of which scored a direct hit on the surface with a missile – or should we say cannon ball?

This periodic comet has had an interesting history after being discovered by Wilhelm Tempel in 1867. Its orbital period was then 5.5 years. After its discovery it was seen again in 1873 and 1879, and was then lost. In 1881 the comet had approached Jupiter and, thanks to the effects of the giant planet's gravity, the period changed to 6.5 years.

At first glance Tempel 1 doesn't look much different from the sungrazing comet that we encountered earlier. Its nucleus measures 4.7 miles (7.6 kilometres) by 3.04 miles (4.89 kilometres), which is quite large for a comet which orbits so close to the Sun.

Deep Impact first visited Tempel 1 early in 2005. As it approached the comet it separated into two portions, the impactor – a 370 kg copper projectile looking rather like a cannon ball – and the flyby craft. The impactor was aimed at the comet and struck its target on January 12, 2005 (actually, the comet ran over the impactor that had been placed in its way, but the distinction seems rather academic), throwing up clouds of dust and producing a bright flash. This flash was due not just to the impact site itself, but also material heated to around 1000 degrees Celsius which was glowing white hot. In total, more than 4000 tonnes of ice must have been excavated by the impact.

Just minutes after the impact the flyby craft passed within 310 miles (500 kilometres) of the comet, but its view was obscured by clouds of dust, preventing it from seeing any crater that was formed (a key part of the attempt to determine the composition and structure of the comet). On the whole, however, the experiment was immensely successful, and photographs reveal that the comet is more dusty and less icy than expected. In fact, it seems that the nucleus of the comet is only loosely held together; think of a dirty snowbank rather than a tightly packed snowball.

This, though, was not the end of the experiment. While Deep Impact left Tempel 1 behind and embarked on an extended mission, NASA was able to recycle another spacecraft. The Stardust probe had originally visited the asteroid 5535 Annefrank, and later sent dust back to Earth that it had collected from the coma of another comet – Wild-2. Now Stardust was put into a new orbit so that it could approach Tempel 1. It passed within a distance of 112 miles (181 kilometres) on February 15, 2011, and was able to identify the crater created by Deep Impact.

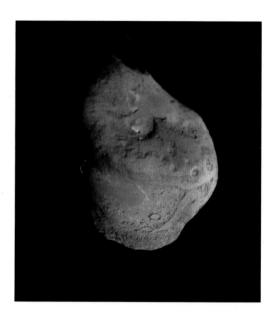

ABOVE: The comet Tempel 1 immediately before the impact, imaged from the Deep Impact impactor.

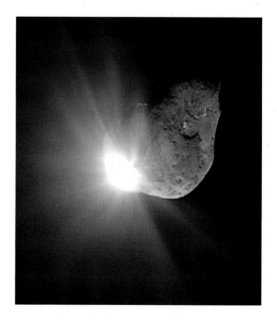

ABOVE: Deep Impact explosion after the impactor had struck Tempel 1.

LAST OF THE ASTEROIDS

Distance from Earth: 35.8 light-minutes

ABOVE: Impression by James Symonds of Thule at the very edge of the asteroid belt.

Before we leave the asteroid belt behind, we have one last stop to make, at an asteroid that has never before been visited by a spacecraft, and there are no plans for a visit in the near future.
Though asteroid 279 Thule is officially included amongst the Main-Belt asteroids, it seems to mark the outer limit of the Main Belt, and there are several curious facts about it.

By asteroid standards 279 Thule is large; its diameter is 79 miles (127 kilometres), probably made up of organic-rich silicates and possibly built around an icy core. This family of asteroids is so unusual that some have suggested that they are interlopers from the outer solar system, rather than anything native to the asteroid belt itself. Its original home may have been closer to the realm of Pluto than anything we've seen on our tour so far.

Wherever Thule came from, its orbit is stable now; the orbital inclination is 2.3 degrees, and its orbit is fairly circular, with an orbital period of 8.84 years. On the far edge of the main group, it has appeared, at least until our visit, as nothing more than a faint dot in the sky. It moves in 4:3 resonance with Jupiter, describing four revolutions around the Sun in the same time Jupiter takes to complete three.

Asteroid families – groups that share similar orbital characteristics and thus, presumably, the same origin – are very common indeed. One of the most common families in the main group is that of the asteroid Hilda. It has been suggested that Thule might be the brightest member of a family, but so far no other member has been found. It may be that Thule moves around the Sun in splendid isolation.

Looking back from Ptolemy toward the Sun, we do not see the array of asteroids and planets in this artist's impression. We see a dark sky with point sources of light from stars, and the occasional bright planet, just as we would see from Earth.

DANGEROUS GIANT

Distance from Earth: 43.7 light-minutes

After leaving the main asteroid belt behind us, Ptolemy takes us into a more or less unpopulated region of the Solar System, in which we encounter nothing more than the occasional comet. Further out lies an unusual cluster of asteroids, not as dense as the Main Belt but numerous all the same: there are believed to be about a million with a diameter of more than half a mile (1 kilometre), about the same number as in the Main Belt. These are the Trojans, moving in the same orbit as Jupiter, but keeping well ahead of it – there is a similar cluster which trails the giant planet as it moves around the Sun. The largest among them, Achilles and Patroclus, are well over 100 miles (160 kilometres) in diameter, but most are much smaller.

The Trojan asteroids, lurking a full sixth of an orbit away from Jupiter, don't have a much better view of the planet (whose gravity has such a profound influence on them), than we do from Earth. We will get a much better view of the planet's cloud tops as we approach, but even from this distance we must be careful; Jupiter is a dangerous place. We must always remember that the planet is surrounded by a zone of radiation that would probably kill any visiting astronaut quite quickly, without the kind of protection offered by our spacecraft. The Earthly equivalent is a region called the Van Allen belts, but they are very puny compared to those generated by the Jovian magnetic field.

One of the first things we notice about the giant Jupiter is the distorted shape of the planet's disc, which is distinctly squashed. The visible surface of Jupiter is not solid, but gaseous, and spinning so rapidly that it is flattened. Despite its enormous size, Jupiter rotates once in less than ten hours. The lack of a solid surface to inspect is perhaps frustrating, but the clouds themselves are mesmerizing.

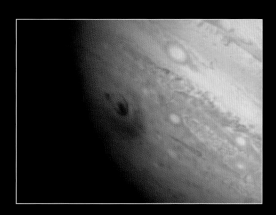

ABOVE: In 1995 comet Shoemaker-Levy 9 crashed into Jupiter's surface. The impact sites of two of the fragments were captured by the Hubble Space Telescope.

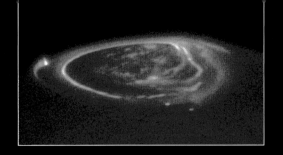

ABOVE: The aurora on Jupiter, powered by electric currents that flow along geomagnetic field lines.

OPPOSITE LOWER LEFT: Aurora at Jupiter's North Pole. This composite image displays X-ray data from Chandra (magenta) and ultraviolet data from the Hubble Space Telescope (blue) overlaid on an optical image.

BOTTOM RIGHT AND MAIN: Cassini spacecraft images of Jupiter. The image at bottom right shows the shadow of the moon Europa.

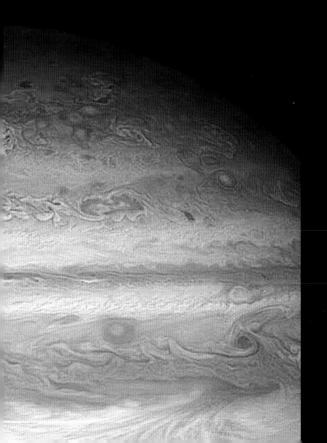

The last cloudy planet we visited was Venus, but the dramatic cloudscapes of Jupiter are very different from the featureless white clouds we saw there. The disc is striped with dark, reddish belts, with a surface that is not stable, but in constant turmoil.

There are two main dark belts, placed on either side of the equator; each is mirrored by other belts, closer to the poles than their more prominent cousins, and the whole system is covered in enormous spots and long chains of clouds known as festoons. The view is surprisingly colourful, with brown belts and reddish spots set against a creamy, pale background. The colours are almost certainly due to the chemical composition of different layers of cloud in the thick atmosphere, but their exact make-up is mysterious. The tops of the Jovian clouds are at about minus 150 degrees Celcius, but the chemicals we think are there shouldn't be brightly coloured when they're that cold.

Whatever the coloured features might be, Jupiter's upper atmosphere is mainly composed of hydrogen. This is the lightest of gases, and small terrestrial planets like the Earth cannot hold on to it any more than the Moon can keep any atmosphere at all. In the early days of the Solar System, though, a proto-Jupiter must have grown large enough to accumulate these gases, accelerating its growth to its present size. In all, the atmosphere is roughly 80 per cent hydrogen and nearly 14 per cent helium, which does not leave room for much else, and the amount of water is negligible. There are, however, appreciable amounts of ammonia and ethane, along with hydrogen sulphide, leading to the inescapable conclusion that Jupiter must be a rather smelly planet, as well as a dangerous one.

Around Jupiter's poles the clouds are darker, but flickering lights can be seen; these are Jupiter's aurorae, the equivalent of Earth's Northern or Southern Lights. Just as on Earth, Jupiter's magnetic fields channel particles towards the poles; but, as the planet's magnetic field, generated deep within the rapidly spinning globe, is the strongest in the Solar System, these aurorae are spectacular.

The aurorae only add to a beautiful and slightly terrifying scene, made all the more impressive by its sheer scale. Many of the spots that lie along the edges of the belts could swallow the Earth whole, and these are storms on a colossal scale. To understand them, we shall take a look at the largest and most famous of them all – the Great Red Spot.

THE GREAT RED SPOT

Distance from Earth: 43.7 light-minutes

The Great Red Spot is Jupiter's most famous feature, and has been present on the surface since at least the seventeenth century.
Associated with one of the southern belts, the spot is enormous, large enough to contain two or three Earths. Despite this great size, the spot is believed to be relatively thin, extending no more than 25 miles (40 kilometres) down into Jupiter's atmosphere. Right now, as we approach it, it is displaying its usual pronounced brick-red colour, though, at times in the past, this colour has faded almost completely.

The Great Red Spot (or GRS) is, in fact, nothing more or less than a giant storm, an anticyclone just like we have in our atmosphere on Earth. Unlike terrestrial anticyclones, though, this storm isn't associated with any mountain or other feature of the Jovian landscape, but floats about 5 miles (8 kilometres) above the normal cloud deck. Inside the spot, material is whirling round at a furious rate, with windspeeds reaching almost 200 miles (320 kilometres) per hour. These speeds are reserved for the edges of the storm; there is relative calm at its centre, which is also cooler than the traumatized regions further out.

The most surprising thing about the Great Red Spot is its longevity. Since observations began it has sometimes faded, or disappeared completely for a while, perhaps because of the temporary emergence of a higher level of bright cloud, but it has always returned. Jupiter's atmosphere is a very turbulent place, however, and even this massive feature may one day disappear forever. Just a few years ago the Great Red Spot collided with, shredded and eventually consumed a similar but smaller, feature known as the South Tropical Little Red Spot (or Little Red Spot to its friends).

How do these storms get started? The Little Red Spot had begun life as a white storm, turning red only a few years before being consumed. A similar beginning was seen for Jupiter's latest tourist attraction, a storm known as Oval BA that has grown to be almost half the size of the Great Red Spot itself. Formed through the collision of three white ovals, themselves small storms, Oval BA began to turn red in 2005 and has begun to rival the Great Red Spot in colour, and in wind speeds. Because clouds at different distances from Jupiter's poles rotate at different speeds, the two giant storms pass each other every three years. Perhaps, one day, the two will merge, producing a spectacular display. For now, the two spots stand out, dancing around each other against the turbulent background of Jupiter's atmosphere.

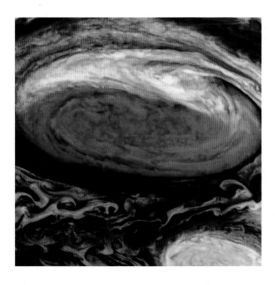

ABOVE: The Great Red Spot in false colours, imaged by Voyager 1.

ABOVE: Hubble image of Oval BA, also known as Red Spot Junior.

OPPOSITE: Jupiter's Great Red Spot as seen by Voyager 1 in 1979.

WORLD OF ICE AND FIRE

Distance from Earth: 43.7 light-minutes

It's not surprising that our stay in this region has so far concentrated on Jupiter itself. The presence of the giant planet is a constant fascination. Tearing ourselves away, though, we soon realize that there is plenty more than just the planet to the Jovian system. The great planet has a large number of moons, and also three rings, each dark and obscure, composed of tiny dust particles, and these rings are visible only when viewing conditions are just right.

The dust is produced by the impact of micrometeorites on some of Jupiter's tiny inner moons, which are nearby, though quite distinct from the rings themselves. Adrastea and Metis form the main ring, and Amalthea and Thebe the even fainter, more tenuous 'Gossamer' rings. These small bodies certainly have their interest, but Jupiter has more than sixty small moons, and the major attractions lie elsewhere, in Jupiter's biggest companions – the four giant satellites that were discovered by Galileo back in the seventeenth century.

The first of the Galilean moons, and closest to Jupiter, is Io, and as we swoop down over it we can see that this isn't a cold, dead world, but one where volcanic eruptions are happening all the time. There are lava flows everywhere, and more than 400 separate volcanoes – one of which is large enough to be visible even in Earthly telescopes, although our view from orbit is much more spectacular. This behemoth, known as Loki, is incredibly violent and we have to be careful to avoid the long plume of smoke that rises from it. The volcanoes emit sulphur, hurling material high above the surface to form a tenuous sulphur dioxide atmosphere and constantly recoating the surface in sulphur frost. This gives Io a spectacular appearance, quite unique in the Solar System, resembling a colourful over-the-top spherical pizza!

Io is a world of contrasts; the general surface temperature is about –140 degrees Celsius, but one eruption of the Pillan Patera volcano, studied by a passing unmanned probe, reached 2000 degrees. From our vantage point in Ptolemy we can see dozens of volcanic vents, spouting geysers and black patches formed by sulphur dioxide frost. It is a beautiful scenario, but also a dangerous one, particularly as Io orbits deep within the deadly Jovian radiation belts.

The source of volcanism in the inner planets is heat left over from the time of their formation, supplemented by radioactive decay, but it might seem that something the size of Io should have cooled long ago. The reason for its continued activity is that its orbit is appreciably eccentric, and as a result, as Io travels round Jupiter, its interior is constantly flexed and heated by the changing gravitational pull. The effect is so dramatic that the shape of Io is distorted by as much as 328 feet (100 metres), between the times when it is closest and furthest from Jupiter, and this tug of war is enough to keep the centre of Io warm, making it a very unusual place indeed.

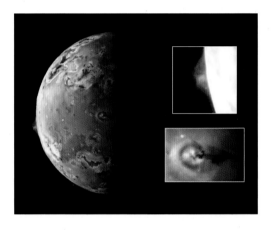

ABOVE: Two sulphurous eruptions are visible on Jupiter's volcanic moon Io in this image from the robotic Galileo spacecraft.

ABOVE: This false-colour picture of Io taken by the Voyager spacecraft shows a volcano erupting on the limb of the Jovian moon.

OPPOSITE TOP: Colour and enhanced images of three different areas of the full disc of Io as viewed by the Galileo spacecraft.

OPPOSITE BOTTOM: Numerous volcanic calderae and lava flows are visible on this volcanic plain on Io. Loki Patera, an active lava lake, is the prominent shield-shaped black feature.

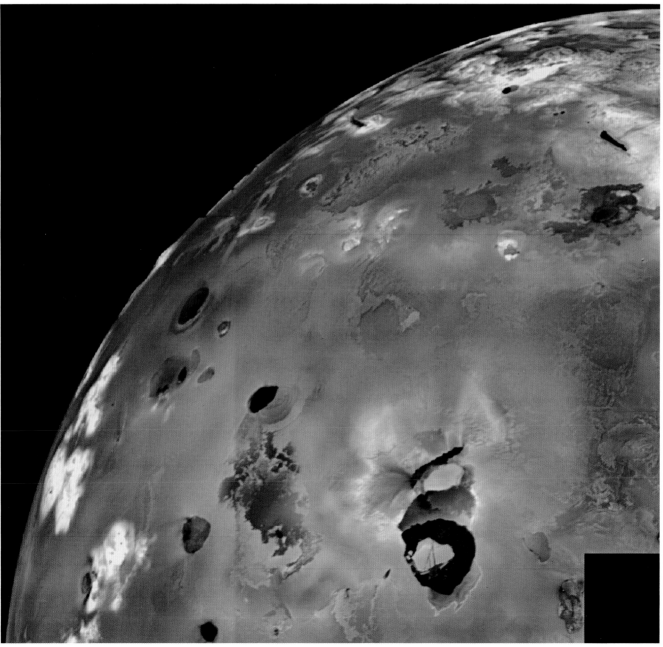

THE SMOOTHEST WORLD

Distance from Earth: 43.7 light-minutes

The second of the Galilean satellites, Europa, is far enough away from Jupiter to be spared the huge volcanic upheaval suffered by Io. As we approach, the surface looks like a sheet of cracked ice, and appearances are not misleading. There are no mountains, no deep valleys, very few impact craters and certainly no volcanoes or lava flows. Europa is one of the smoothest worlds in the Solar System, not dissimilar from the proverbial billiard ball. Let us go closer, and try to make out what Europa is really like.

The ice sheet covers the whole surface, but it gives the impression of being divided into blocks that somewhat resemble the top of icebergs. Strikingly similar formations can be seen in the Antarctic ice shelves back on Earth, and it is believed that here on Europa, just as on Earth, not far below the ice sheet, there may be an ocean of ordinary water. Despite the cold of the outer Solar System, it would be kept liquid by the heat generated in Europa's silicate core. We can do no more than speculate about conditions deep within this inky, dark sea; perhaps one day a mission from Earth will land on Europa and drill through the surface ice. It will not be easy; evidence from the impact craters that are scattered across the surface suggests, despite the best hopes of mission planners, that the ice may be many kilometres thick. Of course, it may be thinner in places, and future explorers might hope that, at least in some regions, it is only a few hundred metres thick.

Will there be life in the Europan ocean? The truth is that we do not know, but it seems at least possible. Some life on Earth has found a way to exist without receiving any energy from the Sun – complex ecosystems exist in the deep oceans, evolving around hydrothermal vents known as 'black smokers' and deriving their energy from the chemicals that are emitted and the heat that is given out. It is even suggested that all life on Earth may have originated in just this sort of environment; and if that could happen on our planet, why not on Europa? Perhaps the first Earthly probe to reach the ocean will encounter the only other complex life forms in the Solar System.

Returning our gaze to the surface, the most striking features are the lineae, dark lines that cross the surface with no apparent regular pattern. These, it seems, mark eruptions of warm ice, as sections of the icy crust of Europa move apart, just as the Earth's crust spreads from mid-ocean ridges. This chaotic network is probably the result of Jupiter's varying gravitational tidal pull, which has an effect on this icy moon after all.

We have time to look only briefly at the other two Galilean satellites as we fly by. Ganymede and Callisto are both ice-coated but covered with craters. Jupiter's system has plenty to intrigue us, but surely nothing can quite rival the cold silence of the sunless sea of Europa.

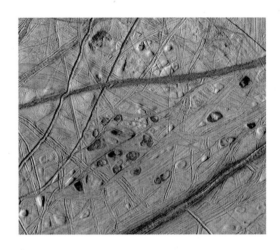

ABOVE: Europa's characteristic surface comprises ridges and cracks. They are seen here along with domes and dark, red spots, or lenticulae - Latin for freckles.

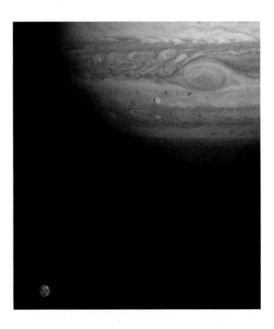

ABOVE: This Cassini image shows two of Jupiter's moons. Europa is the bright moon near to Jupiter's Great Red Spot, while the dark moon in the corner is Callisto.

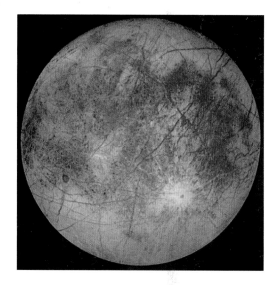

ABOVE: Europa's trailing hemisphere. The prominent crater at lower right is Pwyll and the darker regions are where Europa's primarily water-ice surface has a higher mineral content.

ABOVE: The Hubble Space Telescope took this image of Jupiter's moons in the near-infrared. The three black spots on the face of the planet are the shadows of (left to right) Ganymede, Io and Callisto.

DESTINATION SATURN

Distance from Earth: 1.33 light-hours

As we leave the Jovian system, the rings of Saturn – our next target – are already a stunning sight, just as they are from Earth; but now we won't need a telescope! We will travel many billions of miles on the rest of our journey, and see many rare and unusual sights, but this spectacular scene will be hard to beat!

The rings themselves are composed of chunks of ice, some of them as large as houses and others the size of pebbles. From this distance, the rings give a false impression of being solid, or perhaps liquid, but the truth is that any such ring would quickly be broken up by Saturn's powerful gravitational pull, even if it could have formed in the first place. As we travel toward Saturn, our view of the ring system changes; the rings measure 170,000 miles (274,000 kilometres) from one side to the other, but they are less than a mile (1.6 kilometres) thick, moving in the plane of Saturn's equator.

There are in fact not one, but three main rings: two bright – the A and B rings – and one, the C (or 'Crepe') ring, which is semi-transparent. A and B are separated by the Cassini Division – a small gap of about 3000 miles (4800 kilometres). Despite this basic structure, the whole system is in fact amazingly complicated, each ring being composed of dozens of narrow components separated by thin gaps.

ABOVE: The Cassini spacecraft's view of Saturn.

BELOW LEFT: False-colour view of Saturn's rings by Cassini.

BELOW: Cassini image of Saturn and rings.

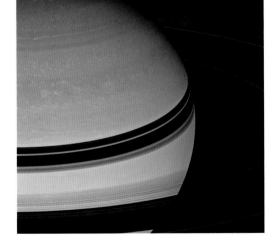

D Ring 74,500 km	C Ring	92,000 km	B Ring

ABOVE: Rhea is in the foreground peeking through the rings of Saturn, with Janus in the background.

BELOW: (left) False-colour Cassini image of Saturn, (right) false-colour rings from Voyager.

BOTTOM OF PAGE: This natural colour mosaic, taken from 10 degrees below the illuminated side of the rings, shows Saturn's rings from left to right, radially outward from Saturn. The total span covers approximately 40,800 miles (65,700 kilometres).

These dark, relatively empty rings are the result of the influence, not of Saturn itself, but of its moons. Take the Cassini Division, for example, which is exactly twice as far from the planet as the moon Mimas, a strange-looking world that resembles the *Star Wars* Death Star. If you were to ride along on a particle orbiting Saturn at the inner edge of the Cassini Division, you would complete two circuits for every one that Mimas managed. This means that the pull of Mimas's gravity is often in the same direction – the two bodies are in resonance – and this rhythmic disruption makes this particle's orbit unstable. Mimas ensures that nothing can remain in the Cassini Division for long, and other moons play a similar role in other parts of the ring.

Saturn's identity is so bound up with its rings – we haven't yet mentioned the planet itself – that it's hard to imagine that they haven't always been here. Yet as the millions upon millions of icy particles orbit the planet, it's equally hard not to wonder about how fragile this whole arrangement really is. Perhaps we're here – as on all the best tours – in exactly the right place at the right time, and the rings are the temporary result of the shredding of a large passing body, perhaps even a comet.

It would have to be a comet since, unlike much of the outer Solar System, the rings are almost entirely made of water ice, but the number of suitable victims and the odds of one passing near Saturn in the last few million years are not known, and there is increasing evidence that the rings formed in the early days of the Solar System.

One attractive theory has a large satellite, approximately the size of today's largest Saturnian Moon, Titan, being ripped apart. The outer icy layers could have been stripped off to form the rings, thus explaining their icy composition, and, as the remaining debris began to clump together it created Saturn's outer icy satellites. The rocky core of the ill-fated moon would have spiralled in, colliding with the planet. The rings themselves would have started off many times more massive than they are today, but gradually material would have been lost, shrinking them to their present size. Perhaps, in a few billion years time, they will be gone.

| 117,580 km | Cassini Division | 122,200 km | A Ring | 136,780 km | F Ring |

SATURN'S SPOKES

Distance from Earth: 1.33 light-hours

Until now we've been viewing the rings more or less as they're seen from Earth, tilted gently towards us. As we manoeuvre up toward Saturn's polar regions we can look down on the rings themselves. This novel perspective makes it easier to see subtle details that further complicate the structure of the rings. In particular, unexpectedly, we can see spokes, dark lines that seem to run from the inside of the rings right out to their edges.

Previous spacecraft passing this way have seen these spokes, too; they may be subtle, but they are not small, stretching up to 6200 miles (10,000 kilometres) in length. The spokes (which come and go over the course of hours, and can appear and disappear very quickly), are composed of the smallest of the ring particles, lifted above the main surface of the rings by electrical charge – in a similar way that a balloon charged by rapid rubbing can cause your hair to stand on end. What exactly is charging the rings isn't clear – suspects include the possibility of impacts by micrometeorites – and the presence of spokes seems very sensitive to the angle that sunlight strikes the rings, making them a seasonal phenomenon.

Let us turn our attention to the planet itself.

ALL: A selection of Cassini images of the spokes in Saturn's rings. Note the straight line in the top image on the opposite page is the very long shadow of one of the moons.

GAS GIANT

Distance from Earth: 1.33 light-hours

As Ptolemy approaches Saturn itself, the planet's appearance, in contrast with the cloudy, turbulent atmosphere of Jupiter, is a smooth, orange surface marked by only a few dark bands. This apparent calm, however, hides a lot of activity; Saturn hosts the fastest winds in the Solar System – speeds of up to 1125 miles per hour (1800 kilometres per hour) have been recorded. These high winds make it very difficult to measure the length of the Saturnian day, but it is now well established at roughly ten and three-quarter hours, extremely rapid for a planet that is the size of more than 760 Earths.

If we switch to viewing the planet in Ptolemy's infrared imager, we will get a very different view of Saturn. The infrared can penetrate the chemical haze that blurs our vision in optical light, and it makes Saturn look as dynamic and interesting as any picture of Jupiter. The most mysterious feature, which has delighted ufologists and other conspiracy theorists since its discovery, is a hexagonal structure that surrounds Saturn's North Pole, each side covering over 8000 miles (12, 875 kilometres).

Its regular features and sharp angles certainly look unusual, but there's nothing artificial about it. Similar features have been created in laboratory experiments which combine a slowly circulating cylinder of water, representing Saturn's atmosphere, with a strong jet stream, and the same phenomena have even been seen in hurricanes on Earth.

The excitement on Saturn isn't always confined to the infrared. Large white storms, composed of icy clouds of ammonia, sometimes disrupt the otherwise placid surface of the planet, creating spectacular trails. Lightning storms are often visible in the upper atmosphere too, occurring most commonly in the summer months, as the Sun stirs up Saturn's atmosphere. This lightning is a sign that these dramatic eruptions are simply thunderstorm piled upon thunderstorm, combining to produce up to ten flashes of lightning per second. The total power in a single storm can match that of the entire planet, making these dramatic events spectacular from up here, but terrifying from down amongst the clouds.

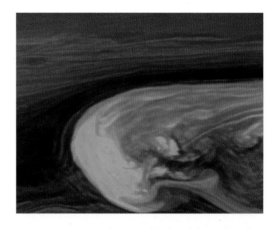

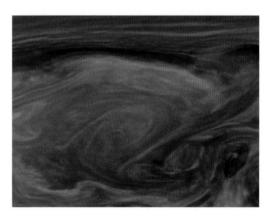

ABOVE: False-colour images of a giant storm circling the planet Saturn recorded in near-infrared by Cassini.

OPPOSITE TOP: Aurorae at the Saturnian pole, recorded by the Hubble Space Telescope in ultraviolet light.

OPPOSITE BOTTOM: Three views of Saturn at different wavelengths; from top to bottom: ultraviolet, visible and infrared.

LEFT: This night-time view of Saturn's North Pole shows a bizarre six-sided hexagon feature encircling the entire North Pole.

IN BLACK AND WHITE

Distance from Earth: 1.33 light-hours

With so many moons around Saturn, one could spend an entire cosmic tour in this neighbourhood alone. Although we have grander plans, we really should visit some of them. We won't get around them all – with sixty-two confirmed satellites not counting ring particles Saturn is almost as well-attended as Jupiter, but we can head toward the outermost member of Saturn's vast family of satellites, Iapetus.

On first impressions, part of the surface of Iapetus is bright and icy, while other regions are completely black. As Ptolemy takes us around this moon we can see that the contrast is really striking, and we come to what may be called the zebra problem – is a zebra a black animal with white stripes or a white animal with black stripes? In the case of Iapetus we know that the density of the globe is very low, so that it must contain a great deal of ice. If ice is Iapetus's natural constituent then it follows that the black areas might be made up of dark material that has been dumped on Iapetus from the surrounding space.

Our best guess is that the material in question might have originally come from Phoebe, the largest of Saturn's outer moons. Indeed, looking carefully, we can see there is a very faint ring of material surrounding its orbit, so perhaps some of this material has found its way to Iapetus, which would pick up dust on its leading face as it travelled through the cloud. In fact, only a tiny amount would be needed; once there, the dust would cause a difference in the absorption of sunlight affecting the ice, and produce the dramatic contrast we see on Iapetus today.

As we fly past, we can see a mountain ridge running along the equator, some 12.4 miles (20 kilometres) wide, 8 miles (13 kilometres) high and 807 miles (1290 kilometres) long. The ridge is a much larger feature than we'd expect on such a small moon; to put that in context, it's much higher than Mount Everest on Earth, and almost as high as Olympus Mons, on Mars. The ridge's origin is unknown, and altogether Iapetus is something of a puzzle.

ABOVE AND BELOW: Infrared images indicate that the dark matter may contain carbon. In this image from Cassini a huge impact crater is evident with a diameter of 280 miles (450 kilometres).

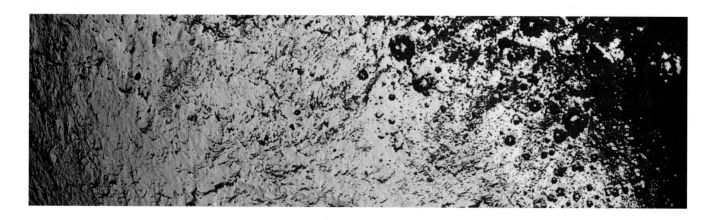

ALIEN LAKES

Distance from Earth: 1.33 light-hours

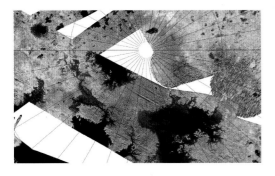

ABOVE: This Cassini false-colour mosaic shows all radar images to date of Titan's north polar region. Liquid hydrocarbon lakes are shown in blue and black. Areas likely to be solid surface are tinted brown.

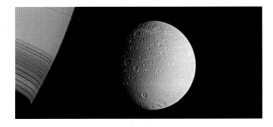

ABOVE: The moon Dione with the rings of Saturn behind, seen from Cassini.

ABOVE: This was the first image Cassini took that revealed a complex geologic surface thought to be composed of icy materials and hydrocarbons. Subsequent images confirmed the existence of lakes of methane.

Looking inward toward Saturn we can see several other moons. One of these, Hyperion, is small and irregular in shape, but some of the others are much more substantial; Rhea, Dione and Tethys are each over 500 miles (800 kilometres) across, with icy, cratered surfaces. But Titan is dominant, and is actually larger than the planet Mercury, though not quite so massive. Clearly it is exceptional, so let's go down and look at it!

Titan, unusually among moons, has its own significant atmosphere, which is dense enough to hide the surface very effectively. We already know it's made up chiefly of nitrogen, with thick methane clouds. Anything could lurk under the clouds, and from here it seems certainly plausible that there might be wide oceans, though these would be seas of liquid methane and other hydrocarbons rather than the water we're more used to experiencing on Earth.

Various unmanned spacecraft passed this way during the twentieth and twenty-first centuries. One of these, Cassini, carried a small lander, Huygens, named after the Dutch astronomer who discovered Titan in 1655. Huygens was designed to cope with landing on an ocean, or on solid ground. In the event, Huygens touched down gently after a somewhat wild ride down through the thick atmosphere, sending back pictures direct from the surface, showing a plain with low hills and what looked remarkably like river beds. The surface layer was found to have the consistency of wet sand, although the icy pebbles that littered the surface were rather more solid – Huygens seems to have hit and cracked one as it touched down.

Some distance away, near Titan's South Pole, we can see what look like lakes. As Ptolemy brings us down we find that they really are lakes, but not of water; these are pools of methane and ethane. Though they are relatively shallow – just a few metres deep – there are other lakes elsewhere on Titan that are so deep that radar waves are not returned, indicating that the lake bottom lies at least 26 feet (8 metres) down.

The light level is low, and clouds hide the sky – there is no chance of a sunny day here. A steady methane rain is falling and it has to be said the scene is not inviting. Certainly there would be no point in trying to fish in the Lake District of Titan! If you are planning a trip, though, it may be as well to remember that methane rain occurs mostly in winter, replenishing the lakes which then seem to dry up throughout the summers.

Titan is fascinating because it seems to offer us a glimpse of what the early Earth may have been like – it is much colder than the Earth ever was, but many of the same ingredients are present. Complex organic chemicals abound, and it may not be too hard to imagine them assembling into more and more complicated configurations, possibly as a precursor to life.

THE FOUNTAINS OF ENCELADUS

Distance from Earth: 1.33 light-hours

Enceladus is a much smaller Saturnian moon than Titan; its diameter is only 320 miles (512 kilometres). We would not expect it to have an atmosphere because its gravitational pull is too weak, and so it was a real surprise when Cassini discovered that an atmosphere did exist – very thin, but quite unmistakable.

More shocks were to follow. On the surface of Enceladus, near the South Pole, there were strange markings which were nicknamed 'tiger stripes'. In Ptolemy we can observe them up close!

The stripes are not mere surface markings; they are deep cracks with steep sides, and water gushes out of them. There are real fountains here, and the crystals that are spewing out into space are made up of water ice, with a small amount of contamination from nitrogen, methane and carbon dioxide. This mix of materials wouldn't be unusual in a cometary jet, but Enceladus is no comet, and it is a long way from the warmth of the inner Solar System, which causes similar activity in comets sweeping through.

THIS PAGE: The colour-enhanced Cassini images of Enceladus backlit by the Sun reveal the fountain-like sources of fine spray. We believe that the jets are geysers erupting from pressurized subsurface reservoirs of liquid water.

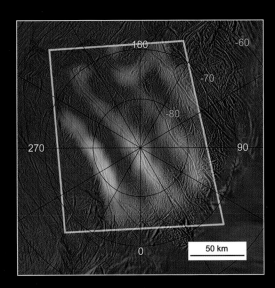

ABOVE: A Cassini image of 'tiger stripes' on the surface of Enceladus.

BELOW: This image shows a number of discrete, fountain-like sources of fine spray.

There seems to be no escape from the conclusion that inside Enceladus there is an ocean of ordinary water that is somehow kept from freezing – partly by the internal heat of the core, and partly because it is salty – salt water does not freeze as easily as fresh water. The closer we fly to the source of the jets, the saltier the composition of the material we can sample. Most of the salt falls back to the tiger stripes, whereas the water ice escapes into Saturnian orbit, where it forms the faint 'E' ring. The salt gives us a critical clue as to the conditions in the ocean; the most likely scenario seems to be that the water is in contact with a small rocky core, the erosion of which produces the salt.

All this seems to make absolutely no sense, bearing in mind Enceladus is so small; Mimas is closer to Saturn, and thus should receive more energy through the kind of tidal forces we found powering activity on Io and Europa, but it is a dead world with a scarred and ancient surface.

There is another problem. Ice crystals are streaming out and might well have been doing so for a very long time. But how could there be enough water to maintain the fountains for millions of years? On the other hand, if the fountains of Enceladus are a more recent phenomenon, then what started them going just in time for our arrival?

We have a ringside view, but frankly Enceladus is a complete mystery. It remains one of the most enigmatic places in the Solar System.

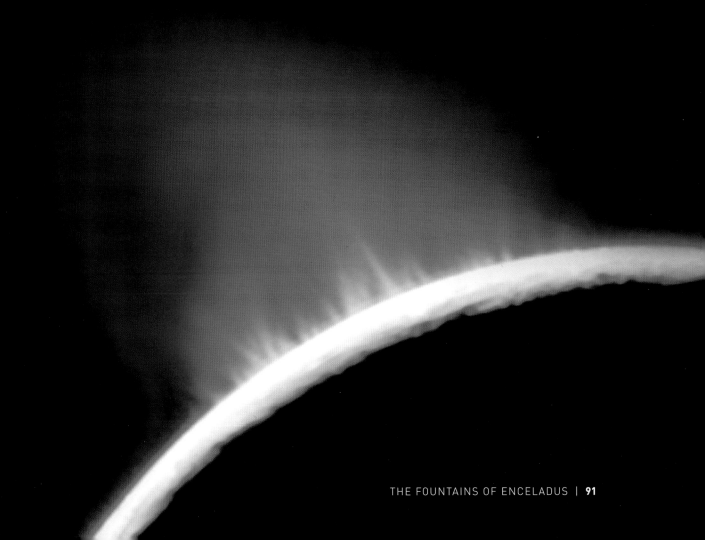

URANUS

Distance from Earth: 2.66 light-hours

Leaving the Saturnian system and the mysteries of Enceladus behind us, let us once more travel outwards, away from the Sun. If we look behind us, our parent star now looks much smaller, and supplies very little heat here in the outer reaches of the Solar System. Since we are about 20 times further away from the Sun than the Earth, the inverse square law that governs radiation tells us we receive only one four-hundredth of the warmth. There is very little else to see before we come to our next destination – the green giant Uranus, named after the first ruler of the Olympian Gods. Just like Saturn, the colour of Uranus is set by the composition of the uppermost layer of clouds, which obscures whatever lies below. Though it still ranks as a giant planet, Uranus is smaller than either Jupiter or Saturn, with a diameter of about 30,000 miles (48,000 kilometres). Ptolemy has now carried us 1783 million miles (2869 million kilometres) since visiting the Sun, and the planet Uranus takes 84 years to orbit it.

Like the other giants, it rotates rapidly, completing one turn in just over 17 hours, but, strangely, it spins in the opposite direction. So if we get Ptolemy to hover above a surface feature, we see the Sun rise in the west and set in the east. This isn't the only strange thing about Uranus's rotation, though. The rotation axes of many of the planets are tilted away from the perpendicular, which is why they have seasons, but Uranus takes this to extremes. The tilt is 97 degrees – more than a right angle – so it seems as if Uranus is rolling along its orbit. How did this happen? An old idea was that the planet must have been hit by a massive body and knocked sideways, but this has always lacked plausibility, and it is now believed that the tilt was increased gradually by interactions with the other giant planets. Uranus's strange spin is just another result of the chaotic beginnings of the Solar System.

Even when we come fairly close to Uranus there are no really prominent markings to be seen. There are ill-defined spots, belts and streaks, but nothing very conspicuous. Uranus appears to be rather bland compared with its larger siblings, but in the Ptolemy, conveniently resistant to all environments, we can plunge into the clouds and dive in toward the centre of Uranus. When we've travelled about four-fifths of the way to its centre, we find that the atmosphere gives way to an icy core. Admittedly, under the immense pressure of the rest of the atmosphere it will behave like a fluid, but this internal structure is different enough from Jupiter and Saturn for Uranus and Neptune to be called not gas giants, but ice giants.

Resurfacing through the clouds we can turn our attention to the dozen or more satellites of Uranus, but all are very small. Probably the most interesting is Miranda, which sports a very varied surface, with craters, cliffs and valleys interspersed with strangely smooth areas. Past geological activity of some sort is probably responsible, but the details remain lost until further study of this desolate place is possible.

ABOVE: Miranda, one of Uranus's 27 known moons, is a jumble of haphazard features: huge fault canyons, terraced layers, and a mixture of young and old surfaces.

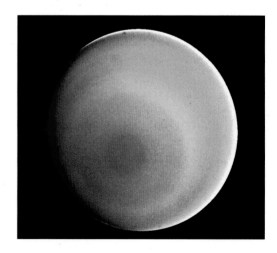

ABOVE: False-colour image of Uranus from Voyager 2.

OPPOSITE BOTTOM: Images of Uranus taken by the Keck Telescope on Mauna Kea, Hawaii. The infrared wavelengths reveal detailed cloud patterns in Uranus's atmosphere.

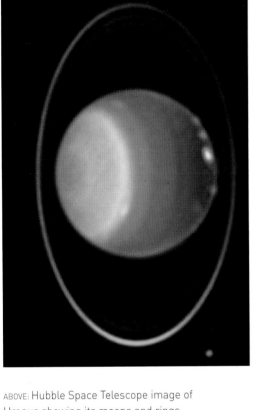

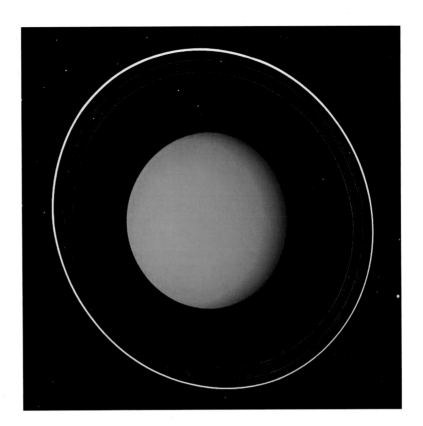

ABOVE: Hubble Space Telescope image of Uranus showing its moons and rings.

ABOVE: Voyager 2 discovered an eighteenth moon for Uranus in 1986. Since then a further 11 have been found.

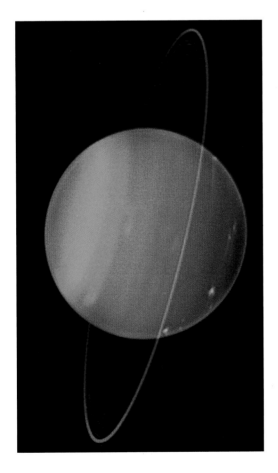

THE OUTERMOST GIANT

Distance from Earth: 4.17 light-hours

To reach our next planet we once again have an immense distance to travel. From Earth, a twentieth-century spacecraft, Voyager 2, took over 12 years to reach Neptune, but using Ptolemy's thought-speed drive we can arrive instantaneously! Neptune's distance from the Sun is 2793 million miles (4494 million kilometres) and it takes 164 years to make one circuit. Neptune is slightly smaller than Uranus, but denser and more massive. It, too, is an ice giant, but Neptune does have a store of internal heat, so that the surface temperature is much the same as Uranus, even though it is much further away. From Neptune the Sun appears little more than an intensely brilliant point of light.

Even when we are some way from Neptune, we can still see there is much more activity than there was on the bland Uranus. The colour is blue, and there are obvious clouds and spots. The two giants may be twins, but they are no more identical twins than Earth and Venus are, and Neptune does not have the same extreme axial tilt.

Images sent back from Voyager 2 showed a huge dark spot on the surface which seems to dominate a vast area. It was called the Great Dark Spot, and was expected to be long-lived, but we can now see absolutely no sign of it. The Great Dark Spot has completely disappeared, although there are other smaller spots here and there; evidently things on Neptune change quickly.

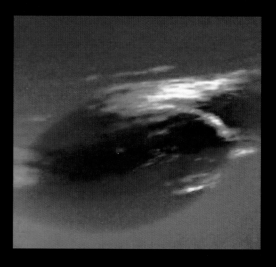

ABOVE: Close-up of the Great Dark Spot from Voyager 2 during its encounter in 1989. The pinwheel (spiral) structure of both the dark boundary and the white cirrus suggest a storm system rotating counterclockwise.

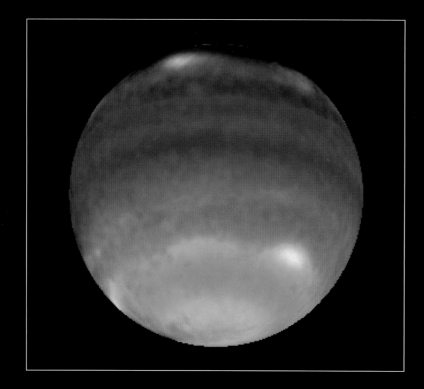

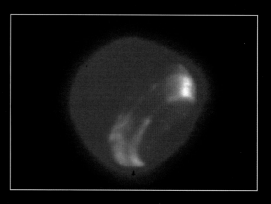

ABOVE: The Keck Telescope's near-infrared images show clouds more brightly than in visible light.

LEFT: The most obvious feature of the planet in this Hubble Space Telescope image is its blue colour, the result of methane in the atmosphere. Clouds are white, and the green belt is where the atmosphere is absorbing blue light.

BELOW: These images were taken by Voyager 2. The pictures show the Great Dark Spot and its companion bright smudge; on the west limb the fast-moving bright feature called Scooter and the Little Dark Spot are visible.

TRITON

Distance from Earth: 4.17 light-hours

Of equal interest to the planet itself is Neptune's one, large satellite, Triton, which moves round the planet in a retrograde sense and is certainly a captured body rather than a bona fide satellite. There is evidence of past cryovulcanism (that is to say icy volcanic activity) along with a ubiquitous coating of icy material, presumably water ice overlaid by ices of nitrogen compounds. Water ice has not been detected spectroscopically, but it must exist, because nitrogen and methane ices are not hard enough to maintain surface relief, not that there is much surface relief on Triton; there are no mountains or deep valleys. Normal craters are scarce. In the Voyager images the most striking feature is the South Polar Cap, which is covered in pink nitrogen snow and ice.

There are also geysers, which were something that had certainly not been expected. Apparently, there is a layer of liquid nitrogen 60–100 feet (20–30 metres) below the surface. Here the pressure is high enough for nitrogen to remain liquefied, but if for any reason it migrates upwards, it will explode in a shower of ice and vapour, and the onrush will sweep dark debris along, which we can see from Earth.

BELOW: The only spacecraft to pass Triton was Voyager 2 in 1989. Voyager found a thin atmosphere and evidence for ice volcanoes.

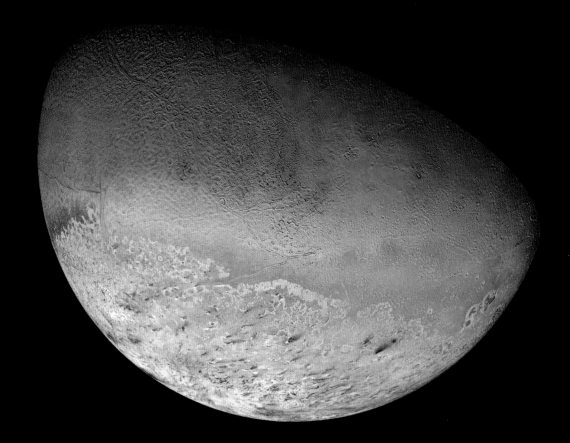

PLUTO

Distance from Earth: 5.64 light-hours

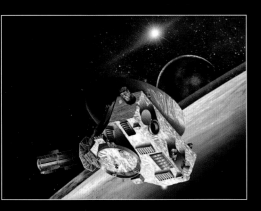

ABOVE: Artist's impression of the New Horizons probe which studied Pluto and its moons in summer 2015 (NASA).

RIGHT: Enhanced colour images of Pluto (lower right) and Charon (upper left), taken by NASA's New Horizons spacecraft as it passed through the Pluto system on 14 July 2015.

Out here in the outer reaches of the Solar System we have to pick our way through a myriad small bodies – the debris which didn't amass to form a larger planet – known as the Kuiper Belt. Amongst the hundreds of thousands of bodies it contains is Pluto. Much smaller than Mercury, Pluto was discovered 60 years before any other Kuiper Belt Object, and so was designated as a planet. But it has now lost its planetary status; it was demoted by the International Astronomical Union in 2006. As a Kuiper Belt Object (KBO), it is not even the largest, or necessarily the most interesting. In 2015 the New Horizons probe provided us with the first ever detailed images of the surface of Pluto.

SEDNA

Distance from Earth: 12 light-hours

Perhaps the most unusual of the Kuiper Belt Objects is Sedna.
Currently 90 times further than the Earth is from the Sun, it has probably the most distinctive orbit of any known object. Sedna is currently swinging inwards towards perihelion (the point at which it is closest to the Sun) in 2076.

Even then, the Sun will seem a very distant object in this lonely world, appearing only one hundred times brighter than the Moon does on Earth. Following perihelion, Sedna will then start a long trek back outwards on its elliptical orbit, taking more than 5000 years to reach its furthest point from the Sun, in a very dark and lonely part of the Solar System. This orbit is so unusual that astronomers struggle to understand how Sedna could have ended up on it; suggestions include disruption from the close passage of a nearby star, perhaps right at the beginning of the Solar System's evolution.

Regardless of how it formed, Sedna is a desolate place today. Its surface is covered by an icy mix of hydrocarbon chemicals, particularly methane. Some of this material will react with what little sunlight there is to form darker chemicals known as 'tholins'. Normally the darkened ice would be churned up by the effect of constant impacts, but Sedna travels a lonely road and, for most of its orbit, impacts are rare, leaving its murky surface intact. This darkness makes Sedna-like objects even harder to find than would otherwise be the case, and there are probably forty or so similar objects lurking undiscovered in the wastes of the outer Solar System. All the more reason to appreciate the one we do know about, as it passes through, on a brief visit to the Kuiper Belt.

ABOVE: The Hubble Space Telescope captured 35 images of Sedna, which shows movement against the stellar background, as you would expect of a Solar System body.

BELOW: An artist's concept of Sedna.

MEETING WITH A WANDERER

Distance from Earth: 12.5 light-hours

ABOVE: Comet Arend-Roland, photographed during its return in 1957. This comet has an unusual 'antitail', a spike of dust sometimes seen as a comet passes through the orbital plane of the Earth.

BELOW: Comet Halley during its visit to the inner Solar System in 1986, seen from the Giotto spacecraft.

Now as we journey out beyond the Kuiper Belt, we come upon something very unusual – an exiled comet, forever banished from the inner Solar System. It is a true lone wanderer in space, and its journey has no obvious ending.

Way back in 1957, when space travel was still regarded as a science fiction dream, a comet was seen in the Earth's sky. It was discovered by two Belgian astronomers, Arend and Roland, and was named after them. It became quite conspicuous, brighter than most of the stars, and had a fairly long tail. It remained visible for weeks, and astronomers using telescopes were able to track it for a year or two, before it receded and was lost to sight.

As a comet recedes from the Sun and becomes colder, it loses its tail and most of the gas around the icy nucleus. Some comets move around the Sun in periods of a few tens of years, so they appear regularly and we always know when to expect them (Halley's Comet is the most famous example with a period of 75–6 years); while others have longer periods amounting to a few centuries, or even tens of centuries and their return cannot be predicted.

Comet Arend-Roland now does not belong to either class. In its journey out from the Sun it passed fairly close to the giant planet Jupiter, and Jupiter's immense pull of gravity threw the comet into an orbit from which it will never return to the vicinity of Earth. Looking out of the windows of our ship, the trusty Ptolemy, it is this luckless comet we have found – moving outwards, ever outwards, away from our Sun and its family; we have no idea what its future will be.

Soon we have left the Kuiper Belt behind us, and there is an immense distance to travel before we arrive at another region inhabited by small bodies; it is the Oort Cloud, at least one light-year from the Sun.

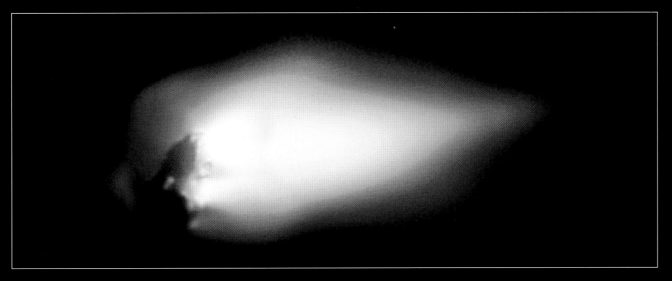

THE LONGEST JOURNEY

Distance from Earth: 18 light-hours

The Ptolemy is not the only traveller in these remote regions of the Solar System. A small craft keeps us company, no larger than a family car and travelling outwards on a path that began with lift-off from the Earth, way back in 1977. This is Voyager 1, until now the most remote man-made object in the Universe, and on board is an ancient form of recording – a gold phonograph record – containing sounds and images selected to portray the diversity of life and culture on Earth. The intent was partly symbolic, but the record is also meant as a message for any intelligent, extraterrestrial life-form, or future human travellers, who may find them. Whether they'll find a suitable record player is another matter!

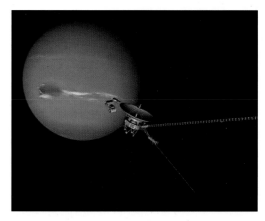

ABOVE: Artist's impression of the Voyager 2 probe at Neptune.

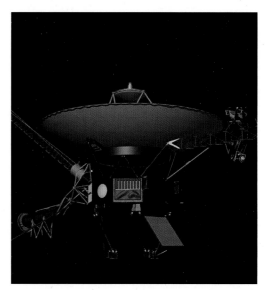

ABOVE: The Voyager spacecraft.

LEFT: A Titan 3E Centaur rocket launching Voyager 2 in August 1977.

OPPOSITE BOTTOM: A composite of the moons of Saturn visited by the Voyager probes.

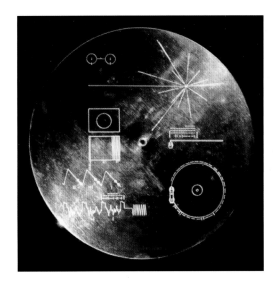

ABOVE: The golden record which was included in Voyager's payload to communicate information to any extraterrestrials about our civilization.

Voyager 1's sister craft, Voyager 2, has travelled more than 9 billion miles (14.4 billion kilometres) at an average speed of 9 miles (14.4 kilometres) a second, or 32,000 miles (51,500 kilometres) an hour. It has visited a string of planets – Jupiter, Saturn, Uranus and Neptune – making it the original cosmic tourist, and is now passing through the region on the edge of the Solar System known as the heliopause, the point where the Sun's influence ceases to dominate.

Most of the instruments and cameras on board the Voyager probes are dead and cold now, deprived of the power they need to operate. A few are still alive, powered by the declining radioactive source that provides the Voyagers with heat and power. The spacecraft are still able to communicate with home via radio, making an incredibly tenuous connection with those who made them and sent them on their way. Even now, though, they are making remarkable discoveries, finding that the heliopause itself has a complex shape. This transition between the region where the Sun's influence is the most powerful thing and the rest of the Galaxy turns out not to be smooth, but to have a complicated shape. Sometime soon – maybe in a month or two, maybe in a few years – Voyager 1 will punch through this transition region, and become the first man-made object to leave the Solar System, an impressive achievement for so small a craft.

LAST STOP IN THE SOLAR SYSTEM

Distance from Earth: 1 light-year

From this distance, the Sun is small, but still much brighter than anything else in the sky and even to the naked eye it retains its familiar distinctive yellow colour. It's hard to believe that the Sun can have much influence out here in the coldness of space, but these remote outskirts of the Solar System are full of activity. We are in the Oort Cloud, a vast reservoir of perhaps a trillion (a million million) small bodies left over from the beginnings of the Solar System, which were scattered out here more than four billion years ago through interaction with Jupiter, and the other embryonic giant planets.

Astronomers on Earth have never seen the Oort cloud; the small asteroids and comets that it contains are far too faint to be detected at this vast distance. Just occasionally, though, a close encounter between two bodies will send one of them flying towards the inner Solar System. Some will shoot in and out, shining briefly as a bright comet; some may sneak by, unnoticed. Others will be captured by the Sun and join the inner Solar System's family, living out their lives as periodic comets.

This replenishing of the inner Solar System happens all the time, but it's believed that sometimes the Oort cloud can be disrupted on a much larger scale. As the Sun moves through the Milky Way, other stars may occasionally come close, and their disruptive gravitational pull scatters the Oort Cloud, sending many thousands of bodies streaming inwards. For now, though, all seems calm and the members of the cloud continue to slowly orbit the Sun, awaiting their turn in the limelight.

We in the Ptolemy, and the Oort cloud, are about one light-year away from the Sun. We have said that the Ptolemy travels at the speed of thought, which means we can skip instantaneously within our Solar System. Compared with our movements, light itself now appears sluggish, taking 8.3 minutes to reach Earth from the Sun, and a few hours to reach the outer planets, but a whole year to reach us out here. So far, we haven't noticed anything unusual happening to Ptolemy's instruments or clocks as a result of our rapid progress, but that's about to change. At this distance, it will now be very apparent that by travelling instantaneously, we are actually travelling back in time, without any assistance from Doctor Who. Looking back at our Sun, and our own blue planet, we are now seeing them as they were a year ago.

Shall we test our theory? Our spaceship Ptolemy is equipped with an extremely sensitive radio receiver. From this position among the icy chunks of the Oort cloud, we can tune into the BBC TV news. Sure enough, we can see that New Year celebrations are still going on. But, they are talking about last year – not this year! It's probably a good time to reflect that travelling at the speed of thought is not something which is currently thought possible.

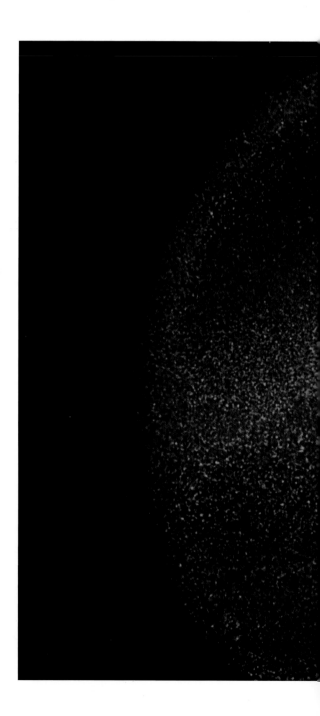

OVERLEAF: A wide-field view of the sky in the region of the planetary nebula Messier 27.

BELOW: Impression of the Oort Cloud by James Symonds, not to scale. From the Oort Cloud the Solar System would appear very small indeed and only the Sun would be visible to the naked eye, as a point of light. One comet has been flung out of the Solar System, and the other is trapped in orbit around the Sun. In reality, at these distances from the Sun neither comet would have a tail – their nuclei would remain frozen.

We are now heading away from our Solar System, toward the constellation of Centaurus; if we linger at a position halfway there, and look back in the direction we've come from, it seems that the familiar 'W' shape of the constellation of Cassiopeia has gained a new star. For all its importance to life back on Earth, the Sun is just that – a star – one amongst many. There is no longer any sign of our home planet, or any other planets; we are saying goodbye to the Sun's family.

Now, turning our gaze outwards, we can see our next destination, Proxima Centauri, which at a distance of four light-years from the Sun is our nearest stellar neighbour. Most of the traditional patterns in the sky are still unchanged – a dramatic reminder that most stars are very far away. The familiar figure of Orion still strides across the sky, since the one light-year we have travelled is nothing compared to the hundreds of light-years that still separate us from these stars.

THE SUN'S NEAREST NEIGHBOUR

Distance from Earth: 4.39 light-years

We are now approaching a star which is very familiar to dwellers on Earth, but only if they live in the southern latitudes. It is Alpha Centauri, the brightest star in the constellation of Centaurus, near the south celestial pole (the point where the imaginary line extended from the Earth's axis meets the stars). We will find that it is, in fact, not a single star, but a complex multiple star system. It's unusual for a star so prominent in the skies of Earth not to have a 'proper' Arabic name of its own; all Alpha Centauri has is 'Rigil Kent', occasionally used by navigators, and an older name – 'Toliman' – which is even more rarely used. Arriving in the vicinity of Alpha Centauri, we first encounter a star somewhat smaller and redder – characteristics which tell us it is cooler – than our Sun. Due to its faintness it was only discovered in 1915, and was christened Proxima Centauri since it is the closest star to our Sun. No planets have been detected moving around Proxima, which is classed as a red dwarf, and it is considered unlikely that any will be found. It is interesting in that it shows sudden flares – outbursts which every now and then make it brighten up very briefly. The flares can grow as large as the star itself.

ABOVE: Proxima seen in the infrared by the UK Schmidt Telescope.

Each star has its own personality, and Proxima's deep-red surface throws an eerie light across Ptolemy's observation deck. Suddenly, though, the light turns orange, and then takes on a more familiar hue – yellow, like our Sun. Proxima's larger siblings, known as Alpha Centauri A and B, have entered our field of view.

The three stars form a single stellar system. A and B are in close orbits around their common centre of mass, each completing one turn every 80 years. Imagine the stars as the twin bells on a spinning dumbbell, orbiting a point on the bar between them, and you'll have roughly the right picture. In actual fact, the orbits are elliptical, not circular, and the distance between the two stars is decreasing rapidly, with their closest approach expected in 2019 when the two stars will be as close together as the Sun and Saturn.

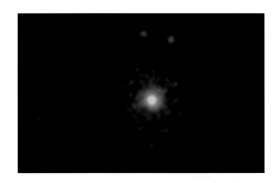

ABOVE: Proxima seen in X-rays, from the Chandra Observatory.

Alpha Centauri B is the smaller and fainter of the pair, its orange tint indicating that it's slightly smaller than our own Sun. Alpha Centauri A, on the other hand, is a close cousin of the Sun – slightly larger, and therefore slightly more luminous, but otherwise indistinguishable – it even turns on its axis at about the same speed as the Sun, rotating once every 22 days. This familial resemblance is an early indication that the Sun is not the special star it seemed at home, but actually a fairly run-of-the-mill denizen of the Milky Way.

Just as A and B orbit each other, so it's believed that Proxima, too, is a true member of this multiple star system. Proxima is further away from the common centre of gravity, and so it moves more slowly. In fact, it's still possible that Proxima isn't really a part of the system at all, but merely an interloper, spending time with its current companions as it passes through space. Only time will tell.

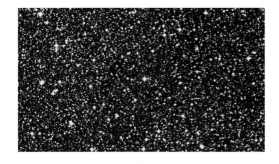

ABOVE: Proxima is the small red star near the centre of this visible-light photograph.

THE DOG STAR

Distance from Earth: 8.59 light-years

ABOVE: Sirius, by Nik Szymanek, above-right of the dome of the William Herschel Telescope.

ABOVE: Sirius is the bright star at top right; the second brightest star in the sky, Canopus, is at top left.

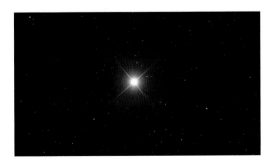

ABOVE: Sirius by Pete Lawrence.

Our next port of call is to something much more spectacular than a tiny red dwarf – we will pay a visit to Sirius, which, viewed from the Earth, shines as the most brilliant star in the sky – where it is easy to find. This is partly because of its brilliance, and partly because Sirius lies in line with the three stars making up the Belt of Orion, a useful signpost for Earthbound astronomers.

Sirius is 26 times more luminous than our Sun, and it is also larger and hotter; its distance from the Earth is 8.6 light-years – about 50 trillion miles. It lies in the constellation of Canis Major, the Great Dog, and so Sirius, itself, is often called the Dog Star. Look at it through a telescope and it appears as a lovely, flashing diamond, glittering with all the colours of the rainbow. But, this is quite misleading, because Sirius is a pure-white star. From the Earth's surface Sirius seems to twinkle, because its light has come to us through a layer of turbulent atmosphere (it's almost as if we were observing from the bottom of a swimming pool), and this shakes the starlight around and breaks it up, to produce the twinkling effect. If you are in a jet aircraft high above the Earth and look out when the sky is dark, you will notice that the stars do not twinkle as much as when viewed from the ground.

Out here, of course, there isn't any air around our spacecraft, and therefore Sirius doesn't twinkle at all. Also, it is now no longer a dot of light but a dazzling blue-white disc, and is throwing out much more energy in the form of hard ultraviolet rays than our own Sun. It is fortunate that Ptolemy screens us from this potentially lethal radiation. Without it we would need eye protection – and sun-screen!

Sirius is not a solitary traveller in space. As Ptolemy brings us in toward it, we can see there is another star, not far away, much fainter than Sirius itself. We have reached another binary system, but this one is not in the least like Alpha Centauri, because its two components are very different in brightness. The Dog Star itself is brilliant while its companion is dim and, because Sirius has been named the Dog Star, the companion has been nicknamed 'the Pup'.

Here we can see it clearly, and this is a very special kind of pup. It is only about the same size as the Earth, but it's incredibly dense – its mass, astonishingly, is as great as the Sun. It is the first and best-known member of a class of stars known as 'white dwarfs'. If we could fill a matchbox with the Pup's material and bring it down to Earth, we would find it would weigh several tonnes. All matter is made up of atoms, and an atom is mainly empty space, but in the material making up a white dwarf the atoms have collapsed under extreme pressure. The protons and electrons are squashed together with very little waste of space, producing a material of amazing density. A white dwarf, such as the Pup, has used up all its nuclear energy, and is near the end of its life; it will eventually lose all its light and heat and become a cold, dead globe.

THE NEAREST PLANET OUTSIDE OUR SOLAR SYSTEM

Distance from Earth: 10.48 light-years

During the last two decades dramatic progress has been made in the hunt for planets elsewhere in the Galaxy – known as exoplanets. Thousands of new worlds are now in the catalogues, and while we can't visit them all, several are particularly interesting.

One that has attracted a lot of attention is in orbit around the star Epsilon Eridani. An orange star, Epsilon Eridani is three-quarters the size of the Sun, but produces only about a third of its luminosity. Its kinship to our own star is at least close enough that it is often considered the nearest Sun-like star, guaranteeing the system a leading role in science fiction. From Ptolemy, the most distinctive features are two belts of debris which surround the star, similar to the Kuiper Belt around the Sun, and one huge planet called Epsilon b.

The planet moves around Epsilon at a distance that is roughly the same as three times the distance of the Earth from the Sun. Being so distant, and orbiting a star that is less powerful than the Sun, it is bound to be a chilly world.

It is not a rocky planet like the Earth. With a mass of at least one and a half times that of Jupiter, it is a gas giant. There will be moons, and many of them will be ice-covered, perhaps even with water, which seems to be a common remnant of the star and planet formation process. Heated not by the faint glow of Epsilon Eridani, but by the internal pull and push of its parent planet's tidal forces, any of these moons might be an Eridanian Europa, with a buried liquid ocean.

Moons or not, the gravitational pull of such a large planet must surely play a large role in maintaining the stability of the debris disc. In fact, computer simulations have shown there may be a few more large planets lurking amongst the asteroids, pulling them this way and that.

During our visit Epsilon Eridani b is easier to see, hanging slightly above the twin dust discs. In fact, the inclination of its orbit has caused some controversy. Earth-bound astronomers cannot see the planet directly as we can (although, by using infrared telescopes, they can pick up the faint glow of the debris discs), but must rely on indirect methods to infer its presence. In the case of Epsilon Eridani b, the discovery came via a technique known as radial velocity measurement, which models the gentle wobbling of the star as the planet orbits it.

Just as the gravity of the star pulls on the planet, so the gravity of the planet pulls on the star. The effect on the star is much less dramatic than that on the planet, but the star will move slightly, as the planet orbits, and this wobble will be detected on Earth as a slight shift in the star's spectrum. A shift to the blue end of the spectrum means the star is approaching, whereas a shift to the red means the star is receding.

ABOVE: Artist's impression of the closest known planetary system to our own, Epsilon Eridani.

HOW FAR THE STARS

Distance from Earth: 11.4 light-years

ABOVE: The Hipparcos satellite measures the positions of stars using the same principles of parallax used by Friedrich Bessel in 1838.

ABOVE: 61 Cygni, Bessel's star.

Looking out at the Galaxy from our spacecraft, despite the vast distances we have travelled, the most remarkable thing is how little the sky has changed; with a few exceptions, the familiar constellations such as the Plough and Orion still hang in the sky.

This is, of course, a consequence of the sheer size of the Milky Way Galaxy. One of the most important scientific discoveries – one which dramatically altered our view of the cosmic scene – was the development of techniques capable of measuring the distances to the stars. The great breakthrough came in the nineteenth century, when several astronomers had the idea of using what we call 'parallax'.

It's easy to explain parallax by means of a simple experiment. Close one eye, and hold up one finger so that it lines up with a distant object, such as a tree in the garden; then, without moving, close this eye, and open the other. Your finger will no longer be lined up with the tree, because you're observing from a slightly different position – your eyes aren't quite in the same place, after all. If you know the distance between your eyes, and can measure the apparent shift in the tree's position – known as the parallax – then you can use simple mathematics to calculate the distance between you and the tree.

Of course, the stars are much too remote to use our eyes in this manner, but the Earth shifts position as it orbits the Sun, giving us an effective distance between our eyes – or baseline – of 186 million miles (300 million kilometres). The astronomers who first tried parallax as a measurement technique selected a star they thought must be close, and measured its position relative to the adjacent stars, first in January and then again, six months later; any shift would reveal the distance to the nearby star.

Astronomers had searched in vain for such a parallax for centuries, but finally, in 1838, the German astronomer Friedrich Bessel measured the parallax of 61 Cygni, a binary star, which attracted his attention because it moved comparatively quickly against the background stars. This pair of orange stars, each smaller and cooler than our Sun, turned out to be some 11 light-years away. They are not particularly unusual, nor are they especially interesting, but they will always have a place in astronomical history. Parallax is still used today, most recently by the European Hipparcos satellite (High Precision Parallax Collecting Satellite), which mapped (with incredible precision), the distance to tens of thousands of nearby stars. Other methods allow us to plot the positions of the rest of the Galaxy's stars. It's time to head out into the Milky Way.

A YOUNG EXOPLANET

Distance from Earth: 25 light-years

Whatever the attractions of searching for a duplicate version of our own Solar System, it isn't just Sun-like stars that turn out to have planets. Moving a little further out, we come across a star much more luminous than the Sun which, like Epsilon Eridani, has a planet moving amongst the debris of star formation. The star in the system is Fomalhaut, 25 light-years from Earth, from where it appears in the constellation of the Southern Fish. While far from being the most powerful star in the neighbourhood, Fomalhaut is still an impressive 1,600,000 miles (2,575,000 kilometres) across and 18 times as powerful as the Sun, glowing with a white-hot light, with a surface temperature of more than 7000 degrees Celsius.

Like our previous destination, Fomalhaut is a young star, which began nuclear reactions perhaps only a few hundred million years ago. Its youth means that Fomalhaut is still surrounded, not by asteroid belts, but by a complete debris disc from which at least one planet has formed. This disc, which has a sharp inner edge some 10,000 miles (16,000 kilometres) from the star, is made up of dust, and shines brightly in the infrared – signifying that it is warm and dense and probably an exciting place to be. In 2008 astronomers using a special high-resolution setting on the Hubble Space Telescope's camera, and blocking out the direct light from the star, detected a planet moving through the debris disc itself.

By examining images taken a couple of years apart, it is now even possible to see the planet moving around Fomalhaut, slowly progressing in its 872 year-long orbit. In fact, the presence of the planet had already been suspected; the sharp, inner edge of the dust disc suggested that particles which strayed closer than the critical edge were being swept up by the gravitational influence of a hitherto unseen body. The exact nature of the interaction between the disc and planet remains obscure, as does the nature of the planet itself. Whatever this world is, whether a gas giant a few times larger than Jupiter, or a small, Earth-like world, it will be cold. More than 100 times further from Fomalhaut than the Earth is from the Sun, it must receive almost no energy from its star.

The presence of such a world around a nearby star raises interesting questions. Is it really common for planets to form so far from their parent stars? Are there worlds huddling closer to Fomalhaut which we have yet to discover and if so, how did their formation affect the material left today in the dust disc we can observe? On the other hand, what if warm worlds, like those in our own system, are rare? Discovering a warm world would have major implications for the chances of finding life, or at least life like our own, elsewhere in the Cosmos. It is time to look for warmer planets.

ABOVE: The planet Fomalhaut b was discovered through its movement. The two bright spots are its positions in orbit around the star Fomalhaut observed in 2004 (left) and 2006 (right), respectively.

OPPOSITE: The star Fomalhaut, photographed by Davide de Martin.

ONE OF THE 'HEAVENLY TWINS'

Distance from Earth: 51.6 light-years

The star system of Castor is a convenient stopping-off point on our tour, and offers the visitor something of a study in contrasts. From Earth the naked eye sees only a single star, but as we approach in Ptolemy, we see much more. As we close in on the system we pass a companion, known as Castor C. It is faint, and red in colour, and close inspection reveals that it is actually a double star comprising two roughly equal components. Such double stars are common, a result of the clearly crowded conditions in the regions in which stars form. In fact, single stars such as our Sun may even be in the minority amid the Galaxy.

Turning our attention to the main star, we see that it, too, is a double, and using our craft's telescope we can see that each of these stars are double, themselves. Castor consists, therefore, of six individual stars, all in orbit around their common centre of mass. The brighter stars are white in colour, but the companion of one is brown, like the components of Castor C.

These separate colours tell us something profound about the stars. There is a close relationship between a star's colour and its temperature; the white stars are much hotter than their brown companions. Inbetween lie yellow stars, such as the Sun. The white stars of Castor are hotter than their companions because they are more massive and, as we will see later on our tour, this means they will have shorter lives. Because of these fundamental relationships between parameters, we can map the entire life history of a star from a distance, merely by observing its colour.

The different colours of the six stars in this vicinity, therefore, tell us about their personalities, but do not represent any difference in origin. The stars of Castor must have formed together something like 200 million years ago, and they are now entering into the long, middle age of hydrogen-burning stars. Stars which are steadily converting hydrogen into helium in their cores are known as 'Main Sequence' stars, and a star like the faint partner of Castor C may expect to spend as much as fifteen billion years on the Main Sequence. The Sun has been a Main Sequence star for four billion years or so, and probably has another four billion to go, while the bright giants of Castor A and B will last only a few billion years. To see what happens after that, we will shortly pay a visit to Betelgeux, in Orion. But for now we head for a very different type of star, Algol.

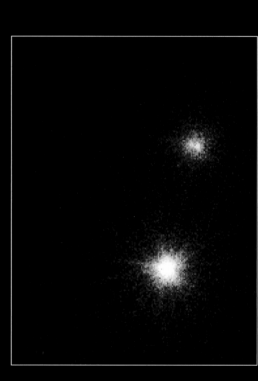

ABOVE: Castor seen in X-rays shows two flaring component stars, while below is the bright star YY Geminorum.

OPPOSITE TOP: Another interesting object in the constellation of Gemini is the open cluster M35.

OPPOSITE BOTTOM: Castor's companion in the constellation Gemini (the Twins) is Pollux, imaged here by Greg Parker.

THE WINKING DEMON

Distance from Earth: 93 light-years

ABOVE: Algol, the brilliant white star, and Rho Persei, the star with the warmer yellow colour.

As we continue outward from the Sun, at around 90 light-years from home, we encounter a most unusual star – Algol, otherwise known as Beta Persei, thanks to its location in the sky in the constellation of Perseus. From a distance Algol looks as impressive as any star we've yet encountered. It is white, and powerful – nearly one hundred times more powerful than our Sun – and almost four times as massive. But as we watch through Ptolemy's portholes, Algol exhibits surprising behaviour. First it begins to fade, and after only ten hours it has decreased in brightness by a full magnitude, appearing less than half as bright as it was at the beginning. Then, after a lull of around twenty minutes, the star slowly begins to brighten again.

After another ten hours it is back to its original brightness, having completed a long, slow wink. This unusual variability has been known since at least the seventeenth century and perhaps before. Its name derives from the Arabic word for ghoul or demon, indicating that ancient astronomers knew there was something noteworthy about it.

Swinging the ship around to get a different view as we approach, the demon reveals the reason for its strange behaviour. Algol is not one star but two, and we see a drop in brightness when one passes in front of the other. This was the first 'eclipsing binary' to be discovered. The 'light curve' (a plot of brightness against time) of an eclipsing binary like this will have a repeated pattern of two dips – a large one when the fainter star passes in front of the brighter one, and a smaller dip when it passes behind. This is indeed observed, but only by those observers who happen to be in a suitable place – in the plane of the rotation of the binary system – which, in the case of Algol, the Earth happens to be. If we take Ptolemy up out of the orbital plane of the system, so we can look down on the two stars of Algol, we see no dips in brightness – just two stars circling a common centre of gravity. There must be many exoplanets missed from Earth because they do not happen to transit their parent star from our point of view, and many Algol-type stars go undetected for the same reason.

We have explained one mystery, but Algol has given its name to another, the 'Algol paradox'. The more massive star is still on the Main Sequence, burning hydrogen into helium, while the less massive of the two has already progressed past this stage, and is a sub-giant which must be converting the helium at its core into heavier elements. This state of affairs makes little sense at first glance, because more massive stars should, thanks to their hotter cores, evolve more rapidly.

The solution is that stars in such close binaries don't evolve independently. The two stars are close enough for something dramatic to have taken place when the more massive star finished burning hydrogen and expanded to become a giant. At that point material was lost to the gravitational pull of the then, smaller star, which then grew at its neighbour's expense, and the pecking order was changed.

A PLANET WITH A TAIL

Distance from Earth: 150 light-years

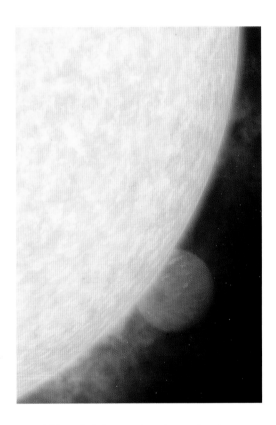

ABOVE: This artist's impression shows planet HD209458b, the first extrasolar planet to have its spectrum detected.

ABOVE: A Hot Jupiter planet, orbiting close to its star in this artist's impression.

HD209458 is an ordinary, Sun-like star. Its yellow light shining on Ptolemy is familiar to those of us brought up on Earth; and just like our own Sun, HD209458 is a slightly variable star, fluctuating in brightness over time, and afflicted with the occasional outbreak of starspots, just as the Sun has sunspots.

One dark shape, silhouetted against the star's bright surface and moving rapidly, is no sunspot – this is the system's major planet, memorably named HD209458b, which completes an orbit once every 3.5 Earth days. The tiny radius of the orbit of this giant planet which lies just 4.3 million miles (7 million kilometres) from its star, makes it an extremely hot world, and there are more dramatic effects yet.

Viewing the planet up close, we can see that it is larger than Jupiter, and against the glare of the star's light, we see a long tail of gas streaming behind it. This gas, mostly hydrogen, accounts for tens of thousands of tonnes of atmosphere lost from the planet every second. The planet's upper atmosphere is boiling away, lost into the vastness of space forever.

This is an extreme example of a class of planets that have become known as 'Hot Jupiters'. While many such worlds have been found close to their parent stars, few are as close as HD209458b and next to none are closer. It's tempting to speculate that the evaporation process that produces this planet's tail has boiled away other bodies completely, destroying planets that stray too close to their stars.

HD209458b's proximity to the star has benefits for us, though. Earth-bound astronomers are a long way from being able to actually image remote planets such as this, but by obtaining spectra of the starlight reflected by the planet, and comparing the intensity of light at various wavelengths with computer models, spectroscopists have been able to estimate the composition of the planet's atmosphere. HD209458b seems to have an atmosphere comparable with that of the giant planets in our own Solar System – a complicated chemical soup which includes atomic hydrogen, carbon and oxygen, along with simple molecules of water vapour, carbon dioxide and methane.

The biggest surprise, though, is how dark the planet appears. Whereas Jupiter reflects more than half the Sun's light that hits its cloud tops, HD209458b reflects proportionally only one-third of light that falls upon it (astronomers call this proportion the object's 'albedo'). This unexpected darkness is, it seems, due to a layer of dark clouds, which themselves separate out the hot hydrogen gas that fuels the tail from the rest of the atmosphere.

This is probably a result of the strong heating which takes place on the side of the planet facing its parent star. Spectacular storms can be driven by this input of energy, with winds produced of more than 4500 miles per hour (7000 kilometres per hour), as material rushes from the hot side to the cooler dark side, driven by the rotation of the planet.

MIRA'S AMAZING TALE

Distance from Earth: 300 light-years

Mira, whose name means the 'wonderful', is an unusual star, the first-recognized truly variable star. It is a binary, but not a particularly close one, and most attention has been lavished on the primary component, Mira A. It is a red giant, and it is wobbling, thanks to instabilities deep in Mira A's core. Once every 10,000 years a particularly violent pulse disrupts the star (lasting something like a decade), and these major pulses have aftershocks that remain visible today. Mira A brightens and fades over a period of 332 days, with the brightening phase taking half as long as the fading phase.

This sort of pattern turns out to be relatively common in stars of Mira's type, and there are now more than 5000 known variable stars in the class that bears its name. The most remarkable feature of the system is, however, only visible in the ultraviolet, which reveals a long and irregular streamer of gas stretching back over more than 15 light-years behind Mira, which is moving through space at a speed of about 80 miles per second (130 kilometres per second). Ahead of the star, the material is shaped into a characteristic bow-shock, flattened before being swept back; this transient feature, a result of Mira's instability and its rapid movement through space, is a remarkable testament to the violence of the forces currently wracking the star.

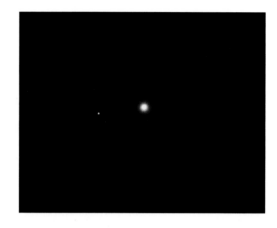

ABOVE: Mira, the red giant star. Its double star is clearly visible.

BELOW: Astronomers recently discovered that Mira has a comet-like tail.

THE SEVEN SISTERS

Distance from Earth: 440 light-years

ABOVE: The Pleiades imaged by Greg Parker.

OVERLEAF: The Pleiades seen from the Palomar Observatory.

BELOW: The Pleiades, photographed by Serge Brunier from Chile.

The well-known Pleiades, or Seven Sisters, are the brightest of hundreds of stars which make up one of our nearest stellar clusters.

The brightest of all the cluster members is a star called Alcyone, and, like the other members of the system, it is a bright blue colour, indicating a luminosity far greater than that of the Sun. The cluster's beauty lies in this particular hue, and is emphasized by the presence of a good deal of 'nebulosity' or cloudiness surrounding the stars. As the stars are young, it has been tempting to regard this background material, which shines only by reflected starlight, as the leftover remnants of the gas from which the stars were formed. However, the evidence we have suggests it could also be possible that the nebulosity is a result of the star cluster travelling through a passing cloud.

The number of stars visible in the Pleiades has also been the cause of much discussion – seven stars can be seen easily with the naked eye from Earth, and keen-sighted people will pick out more; if you can see twelve you are doing really well, and the record is said to be somewhere around eighteen. From Ptolemy's vantage point, hundreds of stars are visible, and they are not all bright and blue.

Amongst the bright Pleiades there are smaller stars, and the smallest of all have come in for intense scrutiny. As the cluster is especially young, it is a good place to catch brown dwarfs, the faintest of stars, before they fade too far below their not-very-spectacular peak luminosity. These brown dwarfs can be as small as a few per cent of the size of the Sun (no more than twenty-five times the mass of Jupiter) and, as you might imagine, they shine very feebly indeed. They are useful tests of our understanding of how stars form, but brown dwarfs also create something of a problem. When is a star too small to be a star? Surveys suggest that the Galaxy is peppered with free-floating objects of this size and smaller, which may even outnumber the stars. Surprisingly, there is no official definition of the minimum size an object needs to be to qualify as a star! The best strategy might be to pick an arbitrary limit – say twenty times the mass of Jupiter – and declare that everything larger than that is a star.

If the estimates of the number of free-floating brown dwarfs are correct, then they must have been expelled from clusters just like the Pleiades. Usually, encounters between stars within the main body of the Galaxy are very rare. The space between the stars is just too great to expect many random encounters between stellar wanderers. But in the high-density environment of a cluster things are different, and stars will often swing past each other. In such encounters double and even triple systems will form and in the vast majority of cases break up, expelling the lighter of the interacting stars. Eventually, the whole cluster will evaporate, with its stars blending into the rest of the galactic throng. This is the fate that awaits the Pleiades, as the cluster slowly disintegrates over the next few hundred million years.

The Exoplanet Kepler 22b

IN THE GOLDILOCKS ZONE

Distance from Earth: 600 light-years

Is there life out here amongst the stars? It's worth keeping an eye out, but where should we look? What sort of conditions are necessary to allow life, and maybe even complex, intelligent beings to form?

All we really know for sure is that those conditions once existed on Earth, but we can tentatively speculate about what was peculiar to this seemingly insignificant third rock from the Sun that allowed the evolution of bacteria and dinosaurs, and even astronomers, TV presenters, and rock musicians.

Life as we know it depends heavily on two constituents – carbon, and water. The chemistry based around carbon (known as 'organic chemistry') is the most complex and creative of any element; carbon's ability to form stable connections to as many as four other atoms, and to form a wide variety of bonds, make it an excellent building block for life. There may be other possibilities, but for now carbon-based life forms seem the most likely inhabitants of the Galaxy.

The case of water is slightly different. It is so fundamental to our life on Earth that it is easy to forget that liquid water is actually a relatively unusual substance in the Universe, bringing together a set of properties that make it extremely useful, and perhaps even essential, for complex life. A solvent is necessary for life, after all – a liquid that will enable the transport of chemicals around the cell, or around the body – and water remains liquid over one of the widest range of temperatures of any chemical (all the way from 0–100 degrees Celsius). The water molecule is relatively abundant in the Universe, formed from nothing but hydrogen and oxygen and, as we shall see later in our exploration, common in star and planet-forming regions.

ABOVE: Artist's impression of Kepler 22b.

There are three other chemical properties that make water an excellent choice for life. Firstly, it is highly 'polar' in that water can form weak bonds with other molecules through electrical attraction and repulsion, producing a wide variety of chemical pathways, of which life – on Earth at least – takes full advantage. Water also has high surface tension, forming drops and pools easily, which leads to such useful properties as the ability to climb capillaries, travelling upwards along a plant's stalk. Finally, and most unusually, when water freezes it expands so that ice floats instead of sinking. If the maintenance of oceans or lakes turns out to be important for the origin and evolution of life, it is much better to have a frozen surface and liquid ocean underneath, than a completely frozen environment.

Most of these attributes are a consequence of the chemistry of water, or at least of the deep quantum physics that determines the molecular properties of the substance. Whether liquid water actually exists, however, will depend upon the temperature of a planet.

At our next destination, the Neptune-sized planet known as Kepler 22b orbits its star, originally called Kepler 22, in what is known as the 'Goldilocks' or 'habitable' zone, a region which is neither too hot nor too cold, but just the right temperature for liquid water to exist.

Kepler 22b was discovered by a different, although still indirect, method from the planets we've visited already. It was found using the Kepler Space Telescope (named after Johannes Kepler, the discoverer of the 'laws' of planetary motions), which stares at a single patch of sky, monitoring the brightness of more than 140,000 individual stars. Just occasionally, a planet will pass in front of one of these stars as seen from Kepler, causing the star to blink for a few minutes or hours. This 'blink', known as a transit, is the signature of a planet and, with enough careful follow-up, can reveal a lot about the planet in question. 22b, for example, orbits its parent star once every 290 days. It is a large planet, with a radius 2.4 times that of the Earth, but we have not yet determined 22b's mass, although we know it is more massive than the Earth, and may be more like Neptune than the Earth. 22b could even be a water world, with a completely liquid surface. More careful observation is needed, but for now this world orbiting an otherwise nondescript Sun-like star is the closest thing to a habitable Earth-like planet we know about – a definite highlight of any tour of the Cosmos.

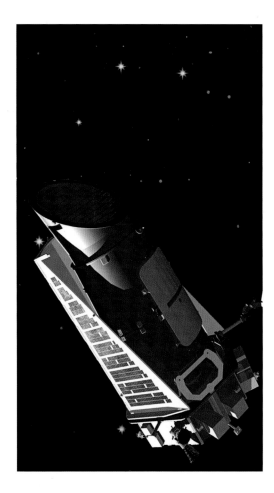

ABOVE: A rendering of the Kepler Space Telescope.

BELOW: Kepler Space Telescope's field of view, superimposed on the Milky Way.

THE COALSACK NEBULA

Distance from Earth: 600 light-years

As we approach, the Coalsack looks rather ominous against its brighter surroundings. A barren void apparently devoid of stars, it stretches out for more than 200 light-years against some of the richest star fields anywhere in this part of the Galaxy. The Coalsack Nebula is large enough to be visible with the naked eye, where it creates a rift in the southern Milky Way. It is not uniform, with some patches revealing a few deep-red stars, and this is the clue to the Coalsack's real nature.

Rather than an empty region of space, the Coalsack is full of dust which blocks out background stars. As we've already seen, astronomers use the term 'dust' to refer to collections of silicate or graphitic grains, each around a tenth of the size of a typical grain of sand. Insubstantial things on their own, when distributed densely enough over a wide area, the dust grains can block out starlight, creating enormous clouds which, protected from the surrounding glare, are amongst the coldest places in the Universe. Such cool temperatures are normally ideal for star formation, and with more than 2500 Suns-worth of mass present, there's certainly enough material hereabouts to produce a substantial cluster of stars; however, the Coalsack seems stable, with no nascent stars embedded within it. Someday, perhaps, a passing star or the ever-changing movements of the Galaxy's spiral arms will disturb it, triggering the beginnings of star formation, but for now it remains dormant, a quiet part of the Milky Way.

Like most 'dark nebulae', as these objects are called, the Coalsack is composed not only of dust, but also of gas. Primarily, this is in the form of molecular hydrogen – difficult to detect from Earth – and carbon monoxide, with a sprinkling of more complicated and interesting molecules. Picking our way through these clouds we can see that the Coalsack isn't one single cloud, but an alignment of two large and several smaller clouds, jostling up against each other. Switching away from optical light, the clouds glow brightly in the infrared, making them distinguishable even from back home. Recent studies with the infrared-sensitive Spitzer Space Telescope have resulted in several of the clouds gaining nicknames according to their apparent shape – one is even known as 'Nessie' as its serpentine shape reminded at least one research group of the Loch Ness monster!

ABOVE: The dark Coalsack Nebula is just below the two very bright stars on the right.

OPPOSITE: The Coalsack Nebula is the dark region in the lower half of the image, and the Jewel Box open cluster is just above centre in the image. The very bright star at top right is Mimosa (Beta Crucis).

A RED SUPERGIANT

Distance from Earth: 640 light-years

Many of the stars we have visited in our tour of nearby planetary systems have been relatively young, but it is time to pay attention to a star at the other end of its lifespan, the behemoth Betelgeux. From Earth this is prominent enough, shining as the second brightest star in the constellation of Orion, where it marks the Hunter's Shoulder. The literal translation of its euphonic name from the Arabic is something like 'the armpit of the great one', but in contemporary astronomy it ends up being pronounced and spelt in several different ways – some people even call it Beetlejuice.

Betelgeux is red, or more accurately orange-red, an indication of the coolness of the star, and its surface is no more than about 3000 degrees Celsius, around the same temperature as a sunspot. Betelgeux makes up for this by being immensely large – its diameter is about a thousand times that of the Sun – and if the two stars were swapped, Betelgeux would swallow up all the planets out to Mars, along with the asteroid belt for good measure. Betelgeux's mass is impressive too; at twenty times that of the Sun, it is a true stellar heavyweight. The most powerful of Earthbound instruments show it as

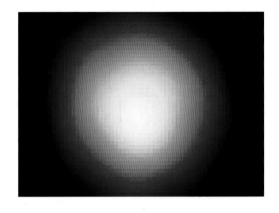

ABOVE: Image of Betelgeux by the Hubble Space Telescope.

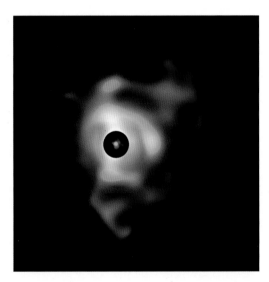

ABOVE: Betelgeux' bright centre is masked by a black disc to allow the material being expelled by the red supergiant to be viewed.

LEFT: Betelgeux imaged by Greg Parker.

ABOVE: Betelgeux is the brilliant orange star in the top left corner in this photograph of the whole of the Orion constellation.

BELOW: Betelgeux by Davide de Martin.

a tiny disc, complete with a fuzzy suggestion of detail on its surface. From our perspective orbiting the star, we can see that what appeared as dark patches on an otherwise nondescript stellar surface are actually convection cells, regions where material is bubbling up from the hotter interior of the star. These starspots are much bigger than any solar spot, taking up a significant fraction of the star's surface, and they are longer-lived, lasting a matter of months before vanishing.

Like most of the largest stars, Betelgeux is decidedly variable; while its range is reasonably restrictive (there is a rough period of five or so years), the fluctuations of the giant star from one day to the next remain essentially unpredictable, the product of chaotic changes within and around the star's core.

Why is Betelgeux so much more impressive than the Sun? It is more massive, and so, paradoxically, it will evolve much more quickly than our own star. While there is more fuel available for nuclear reactions, the larger the star, the greater the gravitational pressure on its core, and thus the faster reactions will proceed. Once a star as large as Betelgeux uses up all the hydrogen in its core, in what could be a matter of just hundreds of millions of years after its birth, disaster will overtake it; power can no longer be generated, and the star will be unable to resist the collapse of its outer layers under their own weight.

Under this collapse the core will shrink and become denser until helium atoms produced in the earlier hydrogen 'burning' phase begin to collide with each other. Like the hydrogen atoms before them, these atoms can begin to combine in nuclear reactions, producing energy which can support the star's outer layers, delaying its collapse. This shift in the mode of energy production has a dramatic effect on the delicate balance of the star, causing the outer layers to expand. For a star as large as Betelgeux, the result is the red supergiant we see.

The Planetary Nebula NGC 7293

PICTURE AT AN EXHIBITION

Distance from Earth: 650 light-years

Before Ptolemy heads deeper into space, we seek out the first example of what is arguably the most beautiful type of celestial object – a planetary nebula. Neither a planet, nor a nebula, the name could not be more misleading. The name arose because to early observers using relatively low-powered telescopes, the nebulae looked a bit like the planets in our Solar System – having a definite size, rather than the pin-pricks of light that are stars. Seen from Earth, stars never have any angular size which can be measured. When we pause at a distance of 650 light-years, ahead of us we see what appears to be a giant eye – unlike anything else we have so far witnessed on our journey. It is the Helix Nebula, and it was created when a star not dissimilar to our own Sun died.

The extraordinary shape and colours of this planetary nebula are the result of the gases shed by the collapse of this star being illuminated by the remnant star, which is now destined to become what is known as a white dwarf. Inside it, nuclear reactions have ceased, but it remains incredibly hot at around 100,000 degrees, and will take a billion years to cool down.

Planetaries are important, though, as well as beautiful, for they can be used to measure the distances to nearby galaxies. The brightness of the glowing oxygen in the ring is directly related to the brightness of the central star which powers it, and can thus be used to calculate the luminosity of the central source. Once we know how powerfully it shines, and how bright it appears in the sky, we can calculate the distance to the planetary, and to the galaxy that hosts it. This kind of trick is performed with all sorts of 'standard candles', including the Cepheid Variables we will encounter on our next stop, allowing astronomers to construct the ladder of distance measurements that enable us to span the observable Universe. Planetary nebulae, when they can be picked out amongst the background light of the stars in their host galaxies, are particularly valuable, because the measurement doesn't depend on the mix of elements in the glowing gas. That, alone, makes them important stops on any tour of the Galaxy.

But we can also enjoy them at a more basic level, for their extraordinary beauty and diversity. They sit in the sky like pictures at an exhibition.

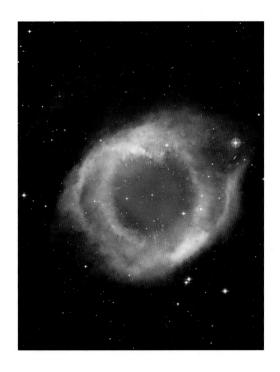

ABOVE: The Helix Nebula seen by the European Southern Observatory in visible light.

OPPOSITE: The Helix Nebula viewed by the Spitzer Space Telescope in the infrared.

COSMIC STANDARD CANDLE

Distance from Earth: 887 light-years

Eight hundred and eighty seven light-years from Earth lies one of the most important stars in the sky – one that had a profound impact on human understanding of the Cosmos. It isn't particularly bright from Earth, and even up close it doesn't look particularly special; in fact, with a surface temperature of around 5500 degrees Celsius, this star – Delta Cephei – isn't far from being solar in appearance, although it is much more massive.

However, ever since it was noticed in the middle of the eighteenth century, astronomers have watched Delta Cephei vary in brightness, getting brighter and then fainter in an extremely regular pattern, taking just over five years to complete a cycle. This remarkable discovery made Delta Cephei one of the first variable stars to be identified, although it soon became clear that it is far from the only star to show this pattern. Nonetheless, it was first, and those that vary in the same way are called Cepheids. Many thousands of Cepheids are now known, but few are as near to the Earth as Delta. One notable exception is Polaris, the northern Pole Star, which, despite recent misbehaviour, still seems to be a Cepheid.

What makes these stars so important is a special relationship between their brightness and the period with which they pulse – the more luminous the star, the longer the period; and because of this correlation, Cepheids came to be used as 'standard candles'. It's possible to estimate the distance to a Cepheid star just by measuring the length of its period of pulsation, so Cepheids have provided us with essential rungs in a cosmic distance ladder that has allowed us to map the Milky Way Galaxy. Moreover, fortunately Cepheids are all relatively bright stars and so can be seen over vast distances, being detected even in other galaxies, so that they can be used to calculate how far away these neighbouring systems are. These very measurements, along with the realization that the light from objects receding from us is reddened (by an effect known as Doppler Shift), enabled Edwin Hubble, in the nineteen-twenties, to establish that the Universe is expanding. Hubble measured the rate at which the galaxies are flying away from each other, as a function of their distance from us, and thus put in place the first building blocks of the modern Big Bang theory. It's a remarkable conclusion, drawn from a series of careful measurements that began with spotting the brightening and fading of an otherwise unremarkable star.

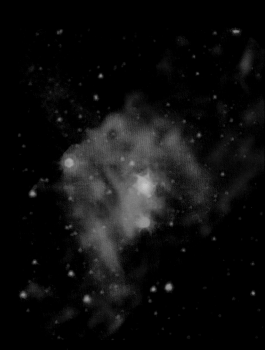

ABOVE: Delta Cephei seen in infrared light. The star has a bow shock in front of it depicted in blue-green, indicative of a wind of gas and dust blowing away the star.

OPPOSITE: This Hubble Space Telescope image of galaxy NGC 3370 is sharp enough to enable us to identify individual Cepheid Variable stars, which have allowed the galaxy's distance to be calculated. In fact, it lies 100 million light-years from Earth.

THE WITCH'S HEAD

Distance from Earth: 900 light-years

We have already visited Betelgeux, one of the two bright stars that do so much to give Orion its distinctive shape in the Earth's skies. It is now time to pay a visit to Rigel, its brilliant neighbour. Rigel is very powerful indeed, shining as brightly as about 40,000 Suns put together, and seventeen times as massive as the Sun. Although small compared to Betelgeux, by most other standards it is very large, with a diameter of sixty-two times that of the Sun – well over 50 million miles (80 million kilometres).

We're not done with Rigel yet, however. Its immense power means that Rigel can have an influence across an enormous range of space, and 200 light-years away we encounter a cloud of gas – a nebula, known as IC 2118 – which is shining thanks to the light of that giant star. This object is also known as the 'Witch's Head' and, looking at images taken from Earth, one can see the resemblance to a cartoon witch, but from our spacecraft the shape is much less clear. The density of the gas in this, as in all nebulae, is much less than the density of the best laboratory vacuum, and its edges are somewhat indistinct.

The nebula is home to a whole series of massive young stars – in particular many of the young and tempestuous types known as 'T Tauri' stars. Most of these stellar teenagers do not shine primarily by nuclear reactions, but by the energy released from their ongoing collapse into their adult form. They are rapidly rotating, and this causes all sorts of activity, from starspots to X-ray flares, and they can easily be distinguished from their more stable older cousins.

The T Tauri stars in the Witch's Head may provide essential clues as to how stars form. We are not sure what begins the process of star formation, but one popular theory is that it is often triggered by previous stars. Initially stars form, and then they disrupt their surroundings through, amongst other methods, the very type of activity we can see in these T Tauri stars; this induces a new round of collapse, and hence a new set of stars, and so on. Of course, one needs to understand what caused the first collapse, but in this case it could easily have been interactions with the powerful light of Rigel, illuminating the side of the cloud nearest to it.

If this simple picture is correct, we should expect the oldest stars to be near the edge, and the youngest stars to be deep within the core of the nebula. Some astronomers think that's exactly what they've seen, but the result remains unclear. Untangling this fundamental mystery of star formation will eventually require the study of many more clouds, just like the Witch's head.

OPPOSITE: The Witch's Head Nebula is powered by Rigel, and its blue glow reflects light from the star. The blue colour is enhanced because dust grains in the nebula reflect blue light more efficiently than red.

AN EARTH-SIZED PLANET

Distance from Earth: 950 light-years

Kepler 22b may be in the Goldilocks zone, but as we saw, it is not an Earth-sized world. To find one of those we have to visit a very different set of worlds, also revealed by the Kepler Space Telescope's unblinking scrutiny. Kepler 20 is a star only slightly cooler than the Sun, surrounded by five separate worlds designated Kepler 20b,c,d,e and f. Other stars are known to have a similarly numerous retinue, and three of Kepler 20's attendants – gas giants which are roughly Neptune-sized – are nothing too far out of the ordinary.

The two remaining planets, Kepler 20e and 20f, however, are very different indeed. When their discovery was announced in December 2011, they became the first unambiguously Earth-sized planets to be found. Their sizes were determined by radial velocity follow-up, using some of the most powerful instruments on Earth, particularly the twin Keck telescopes on Mauna Kea, in Hawaii. One of the planets is approximately Earth-sized, whereas the second is even smaller; together they form a sibling pair strikingly reminiscent of the Earth and Venus.

Planets this small must undoubtedly have a rocky surface, but these are distinctly inhospitable worlds. All five planets in the Kepler 20 system orbit closer to their star than Mercury's orbit round the Sun, so the temperatures on these rocky planets will reach nearly 1000 degrees Celsius – hot enough, perhaps, to melt the surface. Situated much too close to their parent star to be within its Goldilocks zone, it seems unlikely that these roasting worlds in the Kepler 20 system could support any kind of life. They are important, nonetheless, because the five planets tell us that Earth-sized worlds exist, and many more will doubtless be discovered by Kepler and by other efforts in the coming months and years.

Each discovery will tell us more about the rules which govern the chaotic formation of planets. Indeed, there may already be a clue in the arrangement of Kepler 20's planets, which confusingly are arranged so that the small rocky planets each lie between two of the giants. This arrangement, so different from the orthodoxy of our own system which separates rocky from gaseous planets, is rather difficult to explain and suggests that the interactions between forming planets may be rather more complex than we suspected. If so, who knows what surprises lie waiting for us elsewhere?

OPPOSITE: Artist's conception of what Kepler 20e might look like. It is almost certainly wrong, but we won't know the true picture until we arrive.

IN THE SWORD OF ORION

Distance from Earth: 1344 light-years

We have travelled over a thousand light-years from our own Solar System, and encountered a variety of planets around a host of stars, which have ranged from tiny 61 Cygni to enormous Betelgeux. It is now time to see where they came from, as we encounter the first of the great stellar nurseries.

This is the great Orion Nebula, easily visible from Earth with the naked eye (if you're lucky enough to have dark skies), as a misty patch hanging below Orion's belt, 1344 light-years from Earth. The Orion Nebula is our closest large stellar nursery, an immense cloud of primarily hydrogen gas deep within which new stars are being born.

In reality we have been sailing through the nebula for quite some time before we notice it. The gas making up the nebula is incredibly rarefied – millions of times less dense than the air we breathe; its outer limits make their presence known rather indirectly, as its gas and dust block out light at visible wavelengths, making the view of the distant galaxy grow dim. The central regions are, however, lit up by the presence of numerous young stars, particularly four bright stars known as the Trapezium, whose youthful energy not only heats the dust, but energizes the gas, making it shine of its own accord. These four stars lie within about 1.5 light-years of each other. The entire bright portion of the nebula is larger, with a diameter of some 24 light-years, but this is only a small fraction of the total available material.

Where did these stars come from? What makes the Orion Nebula such a good place for star birth? The answers come from studying our surroundings in longer wavelengths than those of visible light. By

LEFT: This infrared view of the Orion Nebula enables us to peer through the dust that obscures the view in visible light.

BELOW: The four immense stars known as the Trapezium power this region of new star formation from the bright yellow heart of this image. Swirls of green indicate hydrogen and sulphur gas, and wisps of red show the presence of carbon-rich molecules.

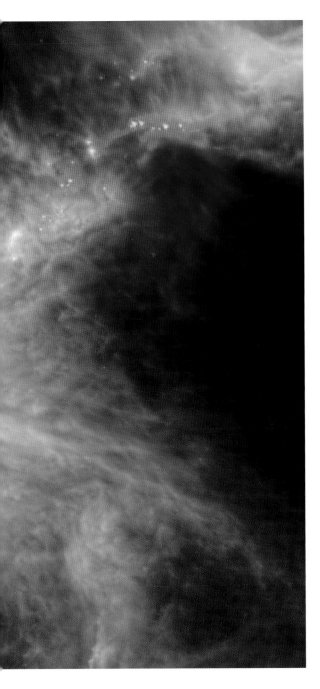

RIGHT: In this nebula adjacent to Orion, with some imagination a reddish figure of a running man can be made out.

shifting our attention beyond the infrared into the submillimetre region (with wavelengths similar to those of radiation generated by domestic microwave ovens), we can see that the nebula is peppered with dark, dusty, cold clumps.

Within these cocoons, the presence of dust blocks out light from the surrounding stars; this allows the material to cool to just a few degrees above absolute zero, typically as low as −270 degrees Celsius. As we've already mentioned, temperature is just a measure of the speed with which a gas's molecules and atoms can move, and at this low temperature they are very lethargic; their mutual gravitational attraction becomes a significant force, able to affect the slow-moving particles, and the clump of gas and dust will begin to contract.

Quite what triggers this process isn't entirely understood, but once begun it gathers pace rapidly. In just a few tens of thousands of years, the density at the core of the clump will have become large enough to enable nuclear reactions to begin, and a star is born. This star will light up its surrounding core from the inside, causing it to glow more brightly in the infrared, and before too long the violent winds, which flow from the as-yet unstable young star, will clear away most of the remaining material. What little is left will form a disc around the star, visible amongst the hotter gas of the surrounding glowing nebula.

The stars of the Trapezium are the most prominent of the nebula's children, but this immense cloud contains a great many young stars which have yet to finish forming. Just like Fomalhaut, they often have discs around them within which material is condensing to form planets. These discs, most easily visible in silhouette against the glowing background gas, are sometimes called propylids; they provide dramatic evidence that newly formed stars retain the raw material that is necessary for the formation of planets.

There is plenty of material left in the Orion Nebula to make more stars, but eventually the supply will run out. The nebula's progeny will dispel the gas around them, leaving large stellar clusters of young stars. We see many such clusters of young stars spread out across the Galaxy, each of which must once have looked like this magnificent stellar nursery. Indeed the Sun and the Solar System must have formed somewhere rather similar more than 4.5 billion years ago.

TESTING EINSTEIN?

Distance from Earth: 1800 light-years

Up to this point on our journey we have been sightseers in the literal sense of the word, exploring the Universe by detecting light, even if we've made use of infrared, ultraviolet and radio wavelengths as well. We are inevitably biased towards objects that emit light, but there are other means of discovery; for example, we can pay attention, not to the light travelling through space, but to space itself.

Since the development of Einstein's theory of relativity, space itself has been an actor on the cosmic stage. It has a substance, and a tension, and is distorted by the presence of matter. In fact, this is how gravity works; it is not too simplistic to say that the Earth orbits the Sun because of the Sun's effect on the surrounding space.

Space (or more strictly, space-time, but the difference need not detain us here) can also ripple, although it is not easy – space is stiff, and it's much harder to get a wave going in it than it would be in an iron girder, for example. Cruising through an otherwise obscure segment of the Galaxy, we can use Ptolemy's sensors to pick up these gravitational waves. This is an incredibly difficult technical task, as even the most significant waves are very small. Attempts have long been made on Earth to try and detect these gravitational waves – the LIGO (Laser Interferometer Gravitational Wave Observatory) experiment, for example, looks for a movement across a baseline of several miles that amounts to not much more than the size of an atomic nucleus!

In September 2015 LIGO detectors measured ripples in the fabric of space-time. Gravitational waves arriving at Earth from the merger of a pair of black holes. The waves we are detecting on Ptolemy, however, have a humbler yet no less interesting source in an object known as the double pulsar.

Here, there are two pulsars, the rapidly spinning remnants of massive stars, in orbit around each other. Their immense gravity powers jets of material which are expelled from their poles; rotating rapidly, these beams sweep through space just as a lighthouse beam sweeps across the coast. Like all pulsars, these are rotating more rapidly than any lighthouse, and we see visible flashes of radiation several times a second. Their rapid movement sets up a series of gravitational waves, which carry energy away from the system, causing the two pulsars to come closer together, increasing the rate of energy loss, and allowing their death spiral to continue. Eventually, the two pulsars will collide, but they are already providing a unique spectacle.

Each of the pulsars acts as an incredibly sensitive clock, with pulses arriving each time the beam of radiation sweeps across our field of view. The clocks are steadily losing time as the pulsars approach each other, and we can calculate the rate at which they must therefore be losing energy. The result agrees exactly with the relativistic equations that predict gravitational waves, making the double pulsar a remarkable test of the most famous theory in physics.

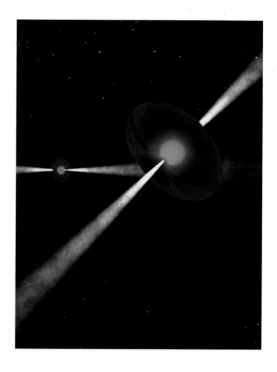

ABOVE: The double pulsar PSR J0737-3039A/B is the only known double pulsar system, with two neutron stars orbiting each other and both visible as radio pulsars; as visualized by Daniel Cantin.

OPPOSITE: The Lovell Radio Telescope, the telescope that discovered the double pulsar.

THE RING NEBULA

Distance from Earth: 2300 light-years

Before we know it, hanging before us is the Ring Nebula, one of the best-known of the planetary nebulae – the beautiful embers of a dying Sun-like star.

Up close the Ring doesn't really look like a planet at all, but has a series of bright shells surrounding a fainter, but still illuminated, central region. Each shell has a distinctive colour, with the edges being red, due to excited hydrogen and nitrogen atoms, and the centre glowing in a pale blue that indicates the presence of extremely tenuous oxygen gas. The ring itself is not smooth, but speckled with denser knots of gas and dust. The whole structure is enormous, more than a light-year across with fainter, ghostly tendrils stretching almost as far out again and visible in the infrared part of the electromagnetic spectrum.

Peering toward the centre of the nebula reveals a faint and yet spectacularly hot star. This white dwarf, glowing with a temperature of about 100,000 degrees, is, in fact, illuminating the whole nebula, and the knots are the result of the interaction of its powerful radiation with the surrounding material. Moving Ptolemy around the object, we can see it is not a ring at all, but more like a cylinder pinched slightly at the waist, around the white dwarf itself.

This is the critical clue to its origin. The Ring Nebula, as with all planetary nebulae, is the remnant of the outer atmosphere of a red giant, shed as the star began to run out of fuel. The core of the star is left as a white dwarf (just like Sirius's companion), and as it is no longer producing energy it is therefore doomed to fade away slowly, over the course of the next few billion years.

The outer layers, which form the beautiful and intricate shapes of the planetary nebulae, are gently ejected in the last few years of the star's life. Their shapes must be guided by the stellar wind, and by the star's magnetic fields which twist and turn the material into their distinctive and unique shapes.

The nebula will disperse in only a few tens of thousands of years. Stars such as this make a critical contribution to the medium around it, spreading heavy elements into the surrounding gas, ready for the next generation of star formation. The Ring Nebula probably formed only 8000 years or so ago, and so we are lucky to be able to stop off at such an unusual, if fleeting, moment in the star's long life.

ABOVE: Seen in the infrared, the Ring Nebula shows looping structures extending far beyond its familiar form; they represent layers of gas and dust expelled by the central star. The central eye is the area shown in the image to the right.

OPPOSITE: We are looking down a cylinder of gas in this image of the Ring Nebula. Dark, elongated lumps of material are embedded in the gas at the edge of the nebula, and the central dying star is in the centre of the blue core.

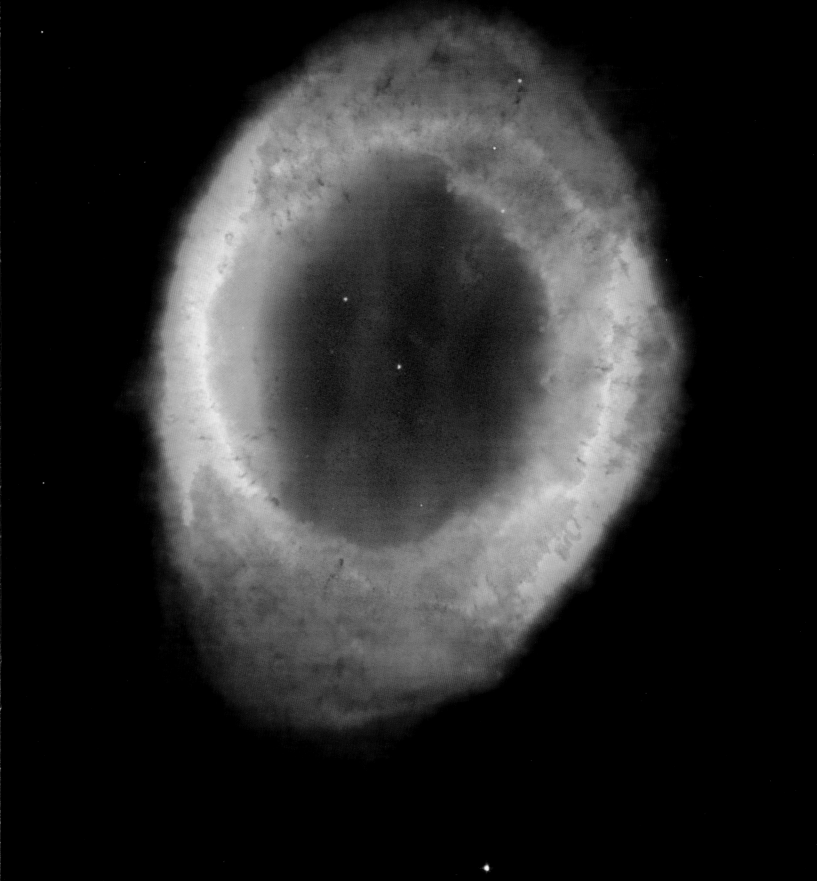

Planetary Nebula HD44179

THE RED RECTANGLE NEBULA

Distance from Earth: 2300 light-years

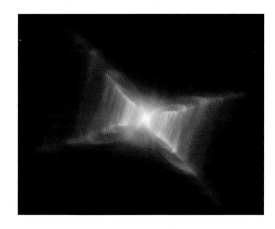

ABOVE AND BELOW: The Red Rectangle is revealed to have more of an X-shaped body in these detailed images from Hubble. The distinct rungs indicate that the outflow occurred in fits and starts.

Our next stop is also a planetary nebula, and here the effects of a binary on the surrounding gas are clear. This nebula is identified by its distinctive colour and shape, and is known as the Red Rectangle, a name which is infinitely preferable to its catalogue identity HD44179.

The nebula is surprisingly symmetrical, with its shape on each side marked by what looks like a ladder of bright regions, with an 'X' marking the central star. Flying around it in Ptolemy shows that the X is merely a side-on view of the same cone shape that we saw at our previous stop; it seems the basic structure of planetary nebulae is always the same, even if the details of individual nebulae are different.

The red colour in the Red Rectangle turns out to be due to the presence in the nebula of complicated molecules, particularly pyrene and a soup of organic molecules known as PAHs (polycyclic aromatic hydrocarbons – on Earth by-products of burning fuels). Organic, remember, means a carbon compound and doesn't have much to do with life. That said, molecules like these indicate a complicated chemistry that might have created the very basic building blocks of life, and which might have been expected to be destroyed in the violent processes that led to the nebula's birth. It seems the PAHs have survived by forming clusters of molecules which are resistant to the influence of light, remaining well into the nebula phase and giving the Red Rectangle its name.

While the Ptolemy is the first rocket to pass this way, it is not the only one to have played an important role in our understanding of the object. The Rectangle was one of hundreds of objects discovered by the first infrared surveys, carried high above the densest part of the atmosphere by rocket flights in the middle of the twentieth century. Such missions are difficult feats of engineering, but without them astronomers can't get access to wavelengths that are otherwise blocked by the absorbent air in our atmosphere. The PAHs we mentioned glow particularly strongly in the infrared, which is why the nebula hadn't been detected by grounded astronomers.

THE ESKIMO NEBULA

Distance from Earth: 2870 light-years

Every planetary nebula is different, and our latest destination – the Eskimo Nebula – is no exception. It owes its name to its supposed appearance through small Earthly telescopes, but the resemblance has never been particularly striking and it is non-existent close up. The nebula appears to have two parts to it, with an inner complex of filaments and bright material surrounded by a dusty ring. The former is probably the ejected material itself, and the latter is marked by strange light-year long filaments pointing away from the still brightly shining central source, clear evidence of interaction between the nebula and the surrounding material.

That central source is again a white dwarf, the remnant of a Sun-like star. It is past the point in its life when it can generate its own energy through fusion, and is now slowly cooling. Such objects, once the planetary nebula around them fade, are difficult to detect, and they must be scattered throughout the Galaxy, doing nothing more than cooling gently over the eons of cosmic time.

BELOW: The Eskimo Nebula, so named because it appears like a person wearing a fur-trimmed parka hood. The filaments are being ejected by a strong wind from the central star.

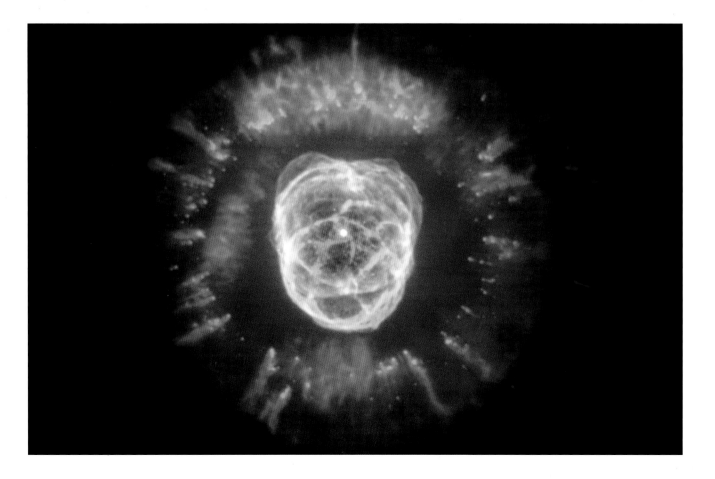

Planetary Nebula NGC 6543

THE CAT'S EYE NEBULA

Distance from Earth: 3300 light-years

Getting used to the extended timescales of astronomy is no easy task. A traveller following in our footsteps in tens of thousands of years' time will find little changed from this edition of the cosmic guide. Some of the stars will have moved appreciably relative to each other, but for the most part they will still shine on, just as they have for billions of years.

The planetary nebulae we have enjoyed are the exceptions; they will last for only a few tens of thousands of years before fading. We really are visiting them at the right time.

From a few tens of light-years away this nebula has an ominous appearance, hanging before us like an eye with a well-defined pupil surrounded by a series of rings. Indeed, its appearance from Earth has led astronomers to give it the nickname 'the Cat's Eye'.

Like the Helix Nebula, it is the dying remnant of a Sun-like star. As we discovered when plunging into its centre, the Sun is powered by atomic collisions that convert hydrogen into helium and burns through four million tonnes of hydrogen a second; it will eventually run out of fuel at its core. When that happens, subsequent stages of stellar evolution will not be quite so stable, and our parent star will eject its outer layers, forming a nebula just like the one we see today.

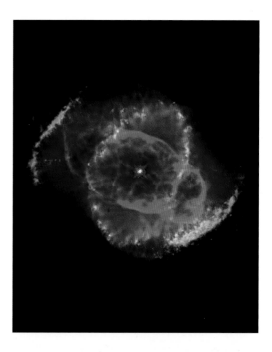

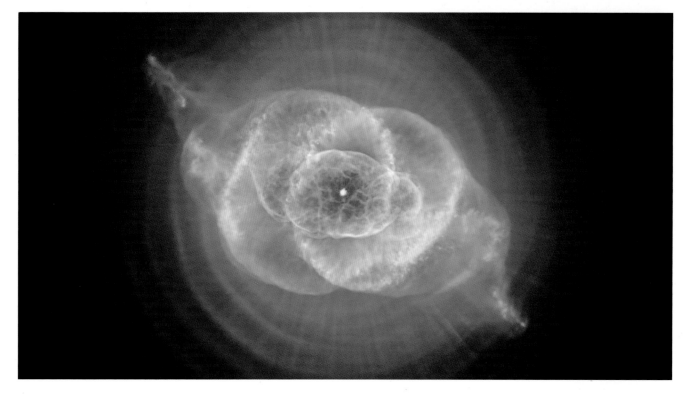

The rings of gas we see, each glowing at temperatures in excess of 7000 degrees Celsius (hotter than the surface of the Sun today), represent a separate violent event in the star's recent past. The central 'pupil' is what remains of the star; it will presumably end its life as a white dwarf, but for now it is hidden from view, and shrouded in mystery. The central source is glowing with energetic X-rays, which may be produced by the accretion of material onto the central star from an otherwise hidden companion.

Such a companion might also account for the complicated structure of the nebula itself. As we manoeuvre Ptolemy around the nebula we can see that its structure is even more chaotic than we suspected; it is made up of a series of interlocking bubbles of different sizes and shapes. The brightest bubble is itself held within a pair of bubbles which join together in an hourglass shape. Such a complicated structure must surely be the result of a complicated process, and the presence of a close binary at the nebula's heart might help, sending the primary star that is responsible for ejecting the material spinning this way and that between outbursts. The star's stellar wind is a powerful influence too, bursting the surrounding bubbles and producing long trails of outflowing material.

Whatever the cause, the show will not last long. The Cat's Eye was likely born no more than one thousand years ago, and planetary nebulae are believed to fade over the course of just tens of thousands of years. The bubbles we can see are still expanding, and the nebula's shape still evolving. Pass this way in a hundred thousand years or so, and there will almost certainly be nothing more than a slowly cooling stellar remnant, with no nebula at all to wonder at.

We have perhaps enjoyed the greatest free exhibition in the Universe, but it is now time to leave the planetaries and head for something a lot less tranquil.

OPPOSITE (TOP AND BOTTOM) AND ABOVE: The Cat's Eye Nebula is one of the most complex planetaries known; possibly there is a binary star at its core. These high resolution images reveal onion-like layers of gas blown off the central star in its latter stages of evolution.

BELOW: the Red Spider Nebula in Sagittarius.

THE CRAB NEBULA

Distance from Earth: 6500 light-years

Nebulae come in many different forms, and among the most unusual of the nearby specimens is the Crab Nebula, otherwise known as M1. From a distance it appears to be nothing more than a patch of light, but, as we approach, we can see an amazingly intricate structure.

The Crab is, in fact, quite unlike any nebula we have seen until now. We have seen nebulae which are places where new stars are being born, and planetary nebulae which are the result of a violent change in a star, but the Crab is the remnant of a star's dramatic death. If we'd been passing this way about 7500 years ago we would have seen a violent explosion. The flash from the explosion was bright enough to be seen in broad daylight back on Earth, some 6500 light-years away, and it was recorded by Chinese astronomers in 1054.

Only the most massive stars end their lives in 'supernovae', as these explosions are called. Having used up hydrogen and then helium in its core, the star will resort to building up heavier and heavier atomic nuclei, burning nitrogen, carbon and oxygen in a series of shells around a core. The last few stages of a massive star's life happen very quickly, but once iron is produced at the core there is no turning back. Iron is the most stable of all nuclei, and colliding two iron atoms together in a nuclear reaction does not produce energy, but uses it up.

A dramatic collapse is therefore inevitable, and the central part of

BELOW: The Crab Nebula seen here in visible light, is filled with complex filaments; green light is hydrogen emission from the exploded star; blue light represents high-energy electrons spiralling in a large-scale magnetic field. The pulsar has been identified with the lower-right of the two stars near the centre of the nebula.

LEFT: The Crab observed from the European Southern Observatory at Paranal, Chile.

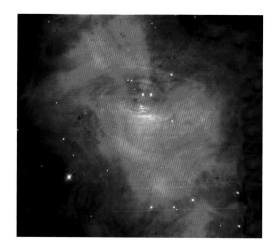

ABOVE: The central portion of the Crab Nebula seen by the Hubble Space Telescope in the middle of the visible part of the electromagnetic spectrum.

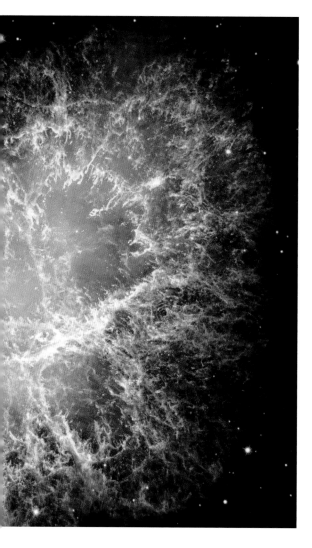

RIGHT: The Crab seen in infrared by the Spitzer Space Telescope.

the star will indeed collapse to form a tiny, heavy remnant. Although only a few miles in diameter in the case of the Crab, the remnant is at least as massive as the Sun. All the atoms making up the centre of the Crab are broken into their subatomic constituents, and are packed together so tightly that even a cupful of this material would weigh thousands of tonnes. It is called a neutron star. Moreover, this tiny remnant is spinning round quickly, rotating thirty times every second, and emitting radio waves as it does so. As this beam of radiation sweeps over our receiver, we pick the waves up as a series of pulses, which are so regular that, when they were discovered, it was seriously contemplated that they might be signals from an alien civilization. The pulses give the name to this type of object – the pulsars we have already met.

Most of the mass of the star does not end up on the pulsar; the collapse of the outer layers of the star is faster than that of the inner layers, and a rebound takes place. It is actually the effect of this rebound or 'bounce' that we see as the supernova, rather than the formation of the pulsar itself. It is these layers, spreading out into the Cosmos, that form the nebula we see today. In fact, the explosion will shred most of the heavy elements that have been produced in the star back down to protons and neutrons, but the density of the ejected material and the energy of the particles are such that further fusion is possible, even during the supernova itself. Indeed, most of the light we see from the supernova is generated indirectly, as a result of the decay of unstable nuclei produced during the explosion.

The material that makes up the Crab will therefore be enriched by elements formed not only in the star, but in the subsequent explosion, which are available for incorporation into future generations of stars. The heavy elements that make up the Earth, you, and this book, were almost certainly formed in an explosion similar to that which produced the Crab Nebula – we are close to this magnificent stellar corpse; this is an awe-inspiring thought.

THE PILLARS OF CREATION

Distance from Earth: 7000 light-years

Our next target, the Eagle Nebula, is 7000 light-years from the Earth; here we find huge pillars of gas and dust, inside which there are hot stars which are in the process of gradually blowing the dust away. Several are visible around the edges of the pillars, just emerging from their surroundings, and a cluster of hot young stars can be seen in the foreground. The scene is so spectacular that it has gained the nickname the 'Pillars of Creation' and these are, beyond doubt, some of the most beautiful objects in our entire Galaxy. The missing portions of the image are a function of the original Hubble camera.

Such dark, dusty nebulae are common, and we have already noted the necessity of dust for the process of star formation. Without it, the molecules of gas which fuel star formation are moving too fast to be corralled by their own gravity, and collapse is difficult, if not impossible. But, where does the dust in the Eagle Nebula come from?

This turns out to be something of a mystery. Massive old stars produce dust in their upper atmospheres and this material can be expelled into the interstellar medium. Supernovae, too, contribute their share of dusty material, but neither process seems productive enough to account for the clouds of dust that we see streaking across the Galaxy. No doubt there is an explanation, but for now the Eagle Nebula's pillars present a puzzle, as well as a stunning vista.

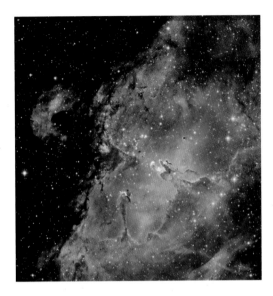

ABOVE AND BELOW: Through this brightly lit window into the Eagle Nebula, we see pillars and globules of gas and dust where stars are forming. Several young blue stars can be seen whose light and winds are pushing away the remaining filaments.

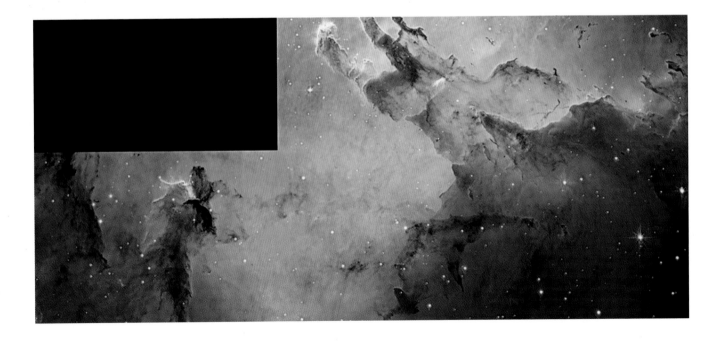

BELOW: This iconic image of pillars of gas and dust in the Eagle Nebula was released in 1995. The pillars are incubators for new stars. Green represents hydrogen, red sulphur, and blue oxygen.

DYING STAR

Distance from Earth: 8000 light-years

We visited the wreck of a star that has blown itself to pieces in the Crab Nebula, but now let us move on to a star that has not quite destroyed itself, yet, but which will certainly do so in the relatively near future. It is generally known by its catalogue designation 'Eta Carinae' even though it does have an old, proper name 'Foramen' – we will refer to it as 'Eta'.

Journeying to Eta takes us to a position about 8000 light-years from Earth, and, as we approach, it's difficult to know what we might find. The star is legendary amongst astronomers because over a century and a half ago – that is to say in the late 1830s when Victoria was Queen of England – it shone as the second brightest star in our Earthly skies, behind only brilliant Sirius. It must have been a stunning sight since Eta shares a patch of sky with Canopus, the second brightest star in the entire sky. Both stars lie in the constellation of Carina, the Keel, too far south to be seen from far northern latitudes, but very prominent in the southern hemisphere. After 1843 it faded, and by now the star is only just visible with the naked eye – a drop of more than a

BELOW: This panorama combines an image around the bright star WR22 in the right half of the picture, with one of Eta Carinae in the heart of the nebula in the brightest region at left. The colourful backdrop of dust and gas in the nebula is illuminated by these massive stars.

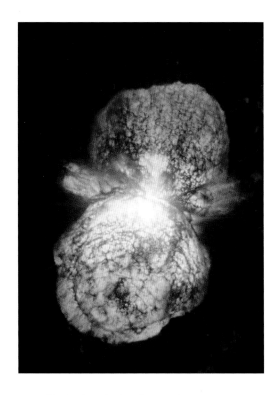

ABOVE: Eta Carinae had a giant outburst 150 years ago, when it became one of the brightest stars in the southern sky. Remarkably, the star survived and produced two massive lobes and a central disc.

hundred times in brightness.

Due to the distance light has to travel from Eta back to Earth, this dramatic brightening actually happened, not in 1830, but 8000 years earlier, in 6170 BC when, back on Earth, the Neolithic era was at its height. By jumping across the Galaxy faster than light, we can see Eta as it will appear from Earth in 8000 years time. Normally, this sort of time period is irrelevantly short, but, for a star that can change as quickly as Eta, it might be important.

The inward route toward the star is not clear, because it is wrapped in a cocoon of swirling nebulosity – primary gas and dust which has been emitted by Eta itself – which is both eruptive and unstable. The star is a remarkable sight, looking more like a glowing dumbbell than any sort of ordinary star; in fact Eta itself is embedded in the two halves of the dumbbell which are the result of that recent 1830 eruption.

Dramatic though it was – and is – the explosion left the star itself relatively untouched. It is a veritable colossus, more than five million times as luminous as the Sun and one hundred or so times more massive, with a diameter not far short of ten million miles (16 million kilometres). Even in its relatively quiescent state, there is nothing calm about Eta; it is restless, with minor outbursts going on all the time, but we know that most of the nuclear fuel has been used up, and that before long energy production will stop – there will be a collapse, an implosion and an unbelievably violent explosion. Eta is preparing to die, but it will depart in a blaze of glory, leaving only a tiny, superdense remnant (almost certainly a black hole). The catastrophe could occur in a million years time (but no longer), or it may happen at any moment – so, beware!

A SWARM OF STARS

Distance from Earth: 15,800 light-years

Ptolemy now takes us out of the main plane of the Milky Way system to a place where we can see the spiral wheeling below us. What we are about to see is a star cluster of a very different nature. The stars in it are not simply scattered around, as they are in clusters such as the Pleiades, but make up a much more symmetrical system. Objects of this kind take their name from this arrangement as they are known as globular clusters.

A globular cluster may contain over a million stars, which become more closely packed toward its centre. Over a hundred of them are known in the Milky Way system alone, but almost exclusively they lie not in the main disc of the Galaxy, but up here in the galactic halo. There is little else here, but the globular clusters stand out, almost as galaxies themselves, slowly orbiting the main disc of the Milky Way. The far-southern Omega Centauri is amongst the best-known and most brilliant of the clusters.

From Earth Omega Centauri is visible with the naked eye, appearing as a blurred patch in the same part of the sky as the Southern Cross, and even small telescopes will show that the outer part is starry; but further in toward the centre the stars are so close together that they cannot be seen separately. At 15,800 light-years from Earth, Omega

BELOW: This image from the Hubble Space Telescope shows 100,000 stars in the core of Omega Centauri. Most stars are yellow like our Sun, older stars towards the end of their life are orange, and red giant stars are red. The faint blue dots are white dwarfs. The bright blue stars are so-called 'blue stragglers' – older stars that acquire a new lease of life by merging with other stars.

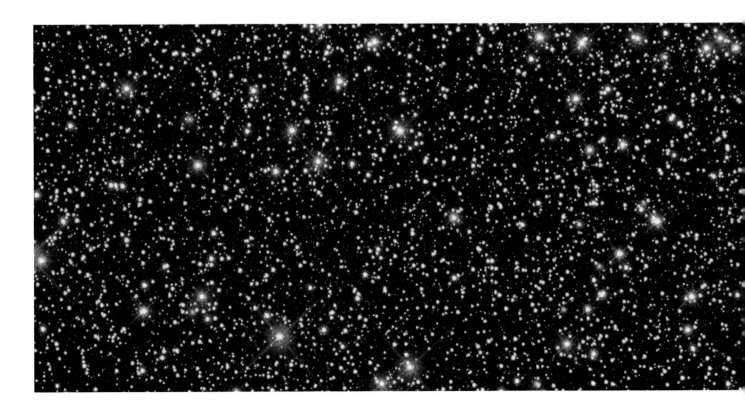

Centauri is actually one of the nearer globular clusters to us.

Once we arrive at the edge of the cluster we realize that from the inner region our view of the Universe will be very restricted. Deep inside the stars really are spectacularly packed together. The whole system is almost 90 light-years in diameter – that is to say, about 540 trillion miles (860 trillion kilometres), and into this area are crammed between one and two million stars. In the central region, the stars are no more than a tenth of a light-year away from each other. As befits the cluster's status as an object that is nearly an independent galaxy, not merely a cluster, there is even a central black hole, although it is not so massive as the one that we will find in the heart of the Milky Way itself.

Looking around, the whole sky is ablaze with many stars much brighter than Sirius, and quite a number are as bright as the Moon appears from Earth. Once we reach the central region, we move in a region of permanent light. There is a strange uniformity to the scene, too. Most of these stars formed together, and we are witnessing the life of a population of stars evolving together, in which, as we have seen, the largest die off first, followed by the smaller members of the cluster's populations.

Are there planets around any of these stars? The high stellar density might be thought to be disruptive, but planetary signs have been observed around stars in the (almost as crowded) centre of the Milky Way. There seems no reason why not, therefore, but any astronomers living there would know little of the outside universe, (their view would be blocked by the glare of starlight from the myriad cluster members), unless they invented radio astronomy, which would then reveal to them the Universe at large. Although Omega Centauri's realm of light is extremely beautiful, it is essentially a backwater, cut off from the main flow of the Universe's evolution.

ABOVE: The view of Omega Centauri from the 2.2-metre Max-Planck Telescope at La Silla in Chile.

THE COSMIC CORKSCREW

Distance from Earth: 18,000 light-years

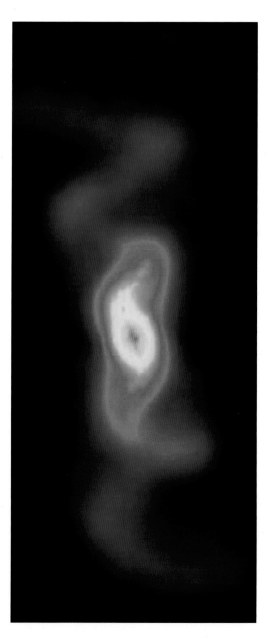

Now let us resume our cruise around our own local spiral arm, in our corner of the Milky Way Galaxy. Within our Galaxy there are about a hundred billion stars, and with so many to choose from there must be few that could claim to be unique. One candidate is the bizarre binary SS433, a massive star in a 13-day orbit around an invisible heavy object.

Although we can't see the star's companion, we can see that material is being pulled away from the visible star, forming an accretion disc around its elusive partner. This disc is dense, and the jostling of material piling on rapidly heats it, so that the disc shines brightly in energetic X-rays. Two large jets of material extend at right angles to the disc, with material moving at faster than a quarter of the speed of light along the jets. If that wasn't strange enough, the entire system is precessing – wobbling – completing a single circle every 162.5 days. This wobble distorts the shape of the jets, leading to SS433's distinctive 'cosmic corkscrew' appearance.

This behaviour can only be explained if the hidden companion to SS433's primary star is a black hole. Presumably, this was once a normal binary, but the largest star must have ended its life in a catastrophic supernova. Indeed, the central object is surrounded by the faint remnant of a supernova, which seems to have occurred something like 10,000 years ago.

The collapse of the core of the very largest stars does not produce a neutron star like that in the Crab Nebula, as the forces that keep particles apart are insufficient to prevent collapse. Instead, an incredibly dense object (with massively strong gravitational pull that even light cannot escape) is formed – a black hole.

The black hole in SS433 weighs between three to thirty times the mass of the Sun. It is a cosmic laboratory; at 18,000 light-years away SS433 is the closest place where we can observe material falling onto the accretion disc around a black hole, and observe the processes that launch jets from the regions close to the central object. The details of the physics that govern this step, and which result in the acceleration of material to such an incredible speed, are not at all well understood, but are critical to an understanding of the evolution of galaxies.

Once material has crossed the point of no return, known as the black hole's 'event horizon' there can be no escape, and it has entered a region of space-time about which we can have very little knowledge. In fact, the so-called 'no hair theorem', widely believed by theorists, suggests that we can know nothing of a black hole beyond its mass, electric charge and angular momentum. The size of the event horizon depends on the mass of the black hole; if the SS433 black hole is three solar masses, then it will be no more than 5.6 miles (9 kilometres) across, which gives a real indication of the immense densities involved.

ABOVE: This radio image reveals the cosmic corkscrew, the jets which are powered by the accretion of material onto the disc around the black hole of SS433. The disc is not visible in this image, but can be seen in other images.

MYSTERY STAR

Distance from Earth: 20,000 light-years

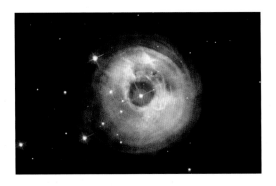

We're a little too late for the most spectacular view of our next stop, a dramatic and unique object which goes by the unprepossessing name of V838 Monocerotis. The star is surrounded by spectacular, swirling dust clouds, illuminated by the light emitted during a sudden explosion which was seen from Earth in 2002. The faint red star at the object's heart suddenly brightened, briefly shining as brightly as a million Suns put together. Its peak lasted only a couple of months before the star rapidly faded and cooled, turning back from a hot, blue star to the faint, red one visible today.

Such a pattern is relatively rare. We see stellar outbursts all the time, and know them by the name 'novae', from the Latin for 'new star', which is how these suddenly bright objects must have appeared to naked-eye observers in the past. The particular path taken by V838 is unusual, however, and it has become the prototype of a new class of 'red novae' which have been detected both in our Galaxy and, perhaps, in a few cases in more distant systems.

Its rareness makes V838 an enticing object for study, as does its unique appearance. The echo of that explosion is still working its way outward today, illuminating the faint dust clouds and casting them into all sorts of improbable shapes. The Hubble Space Telescope has been tracking the progress of these 'light echoes' as they spread outwards from the centre, and astronomers have been keen to study the light in order to work out quite what was going on. Complicating the picture is the presence of a bright companion star, whose light also illuminates and excites the surrounding material, but which has, on occasion, been blocked from view by dense clouds of dust. In fact, right now it's hard to see even from the Ptolemy, hidden amidst a deep, dark cloud which was thrown off in the original explosion.

Untangling the threads of V838's history is thus taking time, but a clear story is beginning to emerge. The nature of the light echo has allowed astronomers to rule out most of the normal causes of novae, most of which are intrinsic to the star itself. Instead, it's believed that a third star, in orbit around V838's now-red central star, got too close and, pulled by its parent's gravity, merged with the larger star. Such stellar mergers are rare; the space between the stars is simply too large for such encounters to happen randomly, and most binary stars are content to remain in stable orbits. When mergers do happen, it seems they create spectacular events, the legacy of which we can see spread out before us in the wonderful and unique sight of V838 Monocerotis.

LEFT: This sequence of pictures dramatically demonstrates the reverberation of light through space caused by an unusual stellar outburst. A burst of light from the bizarre star is spreading into space and reflecting off surrounding shells of dust to reveal a spectacular, multicoloured bull's eye.

JOURNEY TO THE CENTRE OF THE GALAXY

Distance from Earth: 27,000 light-years

We have come a long way in our journey, but we are still inside our Milky Way Galaxy's disc. Before long we will embark on an even longer journey into the depths of the Universe, but first there is one experience that we cannot miss. Over a hundred years ago, Jules Verne wrote his 'Journey to the Centre of the Earth', but we can go further – a journey to the centre of the Galaxy.

It was once thought, with typical human arrogance about our place in the Universe, that the Sun must be the centre of the Milky Way system. Since the time of William Herschel in the eighteenth century, astronomers knew that we lived in a flattened disc of stars, and in 1918 Harlow Shapley realized that we are nowhere near the centre. The true centre is about 27,000 light-years away on the far side of the spectacular star-clouds that fill the constellation of Sagittarius. The whole Galaxy rotates around this centre, and the Sun takes 225 million years to complete one circuit, a period called the 'cosmic year'. One cosmic year ago the dinosaurs were just making their entrance.

One problem with voyaging to the centre of the Galaxy is that our destination cannot be easily seen from Earth, because there is too much material obscuring the way. Visible light, ultraviolet and even

BELOW: This composite image from NASA combines near-infrared images from the Hubble Space Telescope with infrared images from the Spitzer Space Telescope, and X-ray images from the Chandra X-ray Observatory.

low-energy X-rays are blocked, but we can learn a lot from harder, or more energetic, X-rays, gamma-rays, infrared and radio emissions, so that, as we plunge inwards, we have a good idea what to expect.

There is swirling nebulosity, which becomes more and more dense as we journey onward, and the stars become more and more crowded together. We quickly pass by a brilliant cluster, the Arches, made up of about 150 young, hot and intensely luminous blue stars which will shortly explode as supernovae. Another cluster of the same kind, known as the Quintuplet, is also in this region. We are now only one hundred light-years from the Galactic centre, and all the stars are whirling around at a furious rate.

At last we break through, passing the last few stars and being confronted with a disc of material swirling round an enormous black hole, three million times the mass of the Sun. This supermassive beast lurks at the centre of the Galaxy, consuming any material that comes too close. A gas cloud weighing approximately three times as much as the Earth was spotted a few years ago on its way to its doom, and was ripped apart in 2013 when it swung through closest approach (just 3000 times further out than the black hole's event horizon). Stars occasionally pass closer than this, but as they're moving rapidly they escape unharmed; however this particular cloud is unlikely to be so lucky.

Apart from snacking on passing gas clouds, the galactic centre is actually rather quiet. Flares have been seen as material has fallen into the black hole, but in general our Milky Way's black hole is a benign influence, leaving the rest of the Galaxy to go about its business. Stars on their regular orbits are safe, but if we linger we will begin to fall toward the hole, subjecting the Ptolemy to stresses even it cannot withstand. Let us now head up and out of the plane of the Milky Way, and take one more look at our Galaxy from 'above'.

BELOW: An infrared view of the centre of our Galaxy enables us to see deeper into its heart. Clouds of glowing gas and dust, as well as star clusters, can be seen.

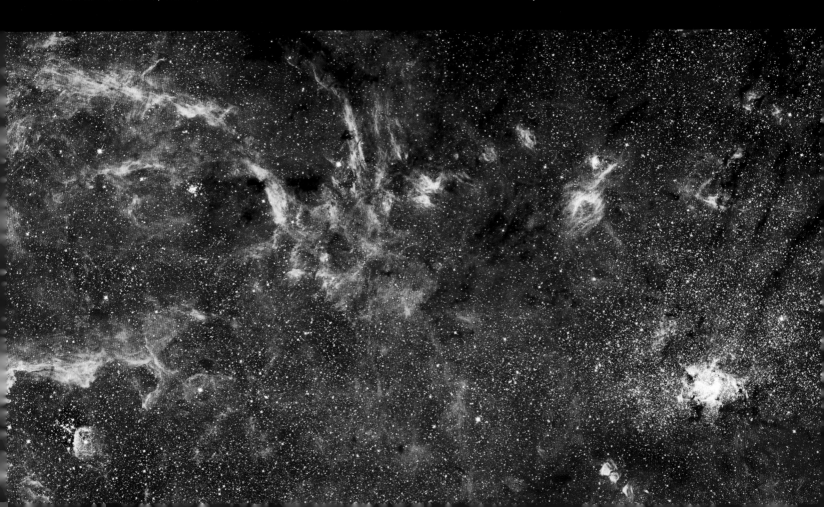

THE MILKY WAY

Distance from Earth: 100,000 light-years

Our trip around the Milky Way is, of necessity, something of a whistle-stop tour, but we should now pause for a moment and look back. From our position, high up above the Milky Way's plane, we have a unique perspective, seeing the structure of our Galaxy laid out beneath us.

Deducing the details of this structure has not been an easy effort for astronomers who are stuck back down in the disc. Large surveys have been undertaken, and the results argued over, a tradition that goes all the way back to William Herschel in the eighteenth century. By making the crude assumption that all stars shine with the same luminosity, and thus interpreting their apparent brightness as a measure of distance, he reached the erroneous conclusion that the Sun was at the centre of a featureless disc, but other astronomers did better and we can see that they drew the main picture just about right.

The structure of the Galaxy is split into the star-forming disc, complete with spiral arms, and a central, yellower bulge containing older stars. The yellow colour reminds us of a fried egg, and that's not a bad mental picture to have; our Milky Way can be thought of as two fried eggs clapped back to back, with the yolks representing the bulge extending above and below the disc of the egg whites.

The nucleus, which we've just visited, is framed on either side by a long bar, from which grow two major spiral arms – named after the

OPPOSITE TOP: The Milky Way as it appears from Earth, seen from La Palma, imaged by Nik Szymanek.

OPPOSITE BOTTOM: The Milky Way seen from the Spitzer Space Telescope at infrared wavelengths. This 360-degree mosaic made up of 2.5 billion pixels shows a one to two-degree slice of the Milky Way. It is like a belt completely encircling us. The centre of the Milky Way is at 0/360-degrees.

BELOW: This picture of the large spiral galaxy NGC 6744 could be a postcard of our own Galaxy – which is very similar.

constellations they pass through from Earth, Perseus and Scutum–Sagittarius. Two minor arms, once thought to be as prominent as their neighbours, lie in between, and are filled mainly with gas and recently active star-forming regions, whereas the major arms have the highest density of both old and new stars. With difficulty, we can pinpoint our Sun's neighbourhood; it turns out that we lie not on a spiral arm, but on a minor feature, the Orion spur, which joins the Sagittarius and Perseus arms. There is also a very faint outer arm, whose status is rather undecided; whether this is a permanent feature or a transient one isn't clear.

The Galaxy is a magnificent sight, and from here set against the blackness of space it seems like a true 'Island Universe'. Yet what we can see of it is only a tiny fraction of the whole. The speed with which it rotates through the long cosmic year may be slow by human standards, but it turns out to be too fast for its gravity to be able to hold it together. In fact, computer models which attempt to explain the rotation only on the basis of the material we see show the Galaxy flying apart in a few million years. Something must be holding the Galaxy together, and we believe the culprit is a mysterious substance called 'dark matter'.

Probably composed of heavy particles which do not interact with light (and which only very rarely interact with normal matter except through gravity), there are good reasons to believe that five-sixths of the stuff in the Universe is dark matter. The visible Galaxy is thus no more than the tip of the iceberg.

Dark matter doesn't form a disc like normal matter, but surrounds the Milky Way's disc in an enormous halo, extending above and below it. The gravitational pull of dark matter keeps the Galaxy together, and may have been responsible for its formation in the first place. If we are right about this, with any luck experiments at the Large Hadron Collider in Geneva and elsewhere will, in the next few years, reveal the true nature of dark matter. Until then, we can only infer its presence by looking at its effect on the rest of the Universe.

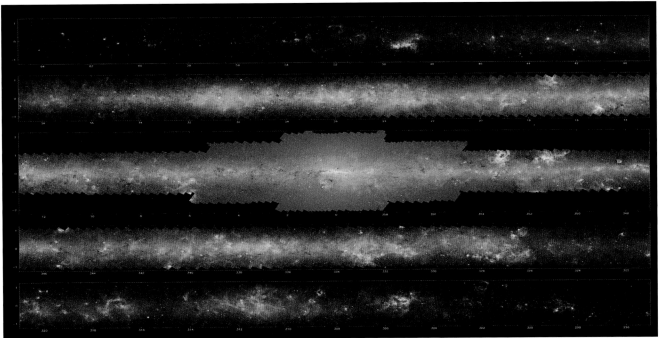

THE MILKY WAY'S COMPANION CLOUDS

Distance from Earth: 160,000 light-years

It is now time to leave our Galaxy. The globular clusters are behind us and we are bound for the great Andromeda Spiral, over two million light-years away. But this is not the nearest neighbouring galaxy, and on the way there we shall pay a quick visit to the two Clouds of Magellan, the largest of the Milky Way's small neighbours.

Against a dark sky these systems look like detached parts of the Milky Way. They are conspicuous in a dark sky, and must have been known since very early times, but the first description we have of them is due to Antonio Pigafatta, a member of Ferdinand Magellan's round-the-world voyage between 1519 and 1522. It may have been fairer to call them the Pigafattic Clouds, but, whatever they're known as, both are satellite galaxies moving round our own, rather like the Moon orbits the Earth.

The large cloud is a journey of about 160,000 light-years away. It may be in orbit around the Milky Way, but the Large Magellanic Cloud is a substantial system in its own right, having about one-tenth the mass of the larger system. From afar we can see it is not a Catherine-wheel spiral, like the Milky Way, though there are some indications of short arms extending from an ill-defined central bar. Its orbit around the Milky Way has often caused disruption; simulations of the last four billion years of the large cloud's orbit show that material is often stripped from the Galaxy when it comes too close to the Milky Way's disc. As it is so much larger than its smaller visitor, the Milky Way is relatively unaffected by these interactions, which may be typical of the way in which larger galaxies fuel themselves.

One particular object of note, toward the edge of the Large Magellanic Cloud, is the remnant of a recent, spectacular supernova. Let's take a closer look.

ABOVE: The dark clouds of dust and gas to the right of this image of part of the Tarantula Nebula look a little like a seahorse, but are in fact areas in which stars are forming.

OPPOSITE: The Large Magellanic Cloud seen in the infrared by the Spitzer Space Telescope reveals nearly one million objects.

BELOW: The Large Magellanic Cloud is composed of a bar of older red stars, clouds of younger blue stars, and a bright-red, star-forming region near the top.

THE GREATEST EXPLOSION OF OUR TIME

Distance from Earth: 167,000 light-years

While amongst the stars of the Large Magellenic Cloud, we should pay homage to one in particular, or rather to its remnant. Amongst the swirling clouds of the small galaxy's nebulae, there lies a series of interlocking rings, each of which is thousands of light-years across.

This is what remains of a blue supergiant which catastrophically ran out of fuel and exploded as a supernova roughly 160,000 years ago. Light from this dramatic event reached Earth in 1987, making this supernova the nearest to be detected since the invention of the telescope, and thus one of the most carefully studied explosions in the history of astronomy.

Astronomers detected not only light from the supernova, but also a burst of particles known as neutrinos which were released in the early stages of the explosion. Careful combing of records definitively identified the star that gave birth to the supernova as that blue supergiant, making SN1987A critical in establishing the truth of the theory that it is the collapse of a massive star that produces these incredibly luminous explosions.

As you might expect, such a supernova can have a profound effect on its surroundings. Although the supernova itself has long since subsided, we can still see the shock it produced forcing its way outwards. It is faint now, but in its earlier days the passage of this shock lit up the surrounding material like a string of pearls, carefully observed by the Hubble Space Telescope back in Earth orbit. There are traces of a turbulent past even before the explosion, too, with fainter outer rings of material surrounding the main remnant. These rings were probably ejected from the star tens of thousands of years before the main event, but have been further distorted by the impact of material from the supernova itself.

With all of this activity stirring up the surrounding interstellar medium, it's possible that the death of the star that produced 1987A will lead to the formation of new stars, their collapse and ignition being triggered by the turbulence it induced. Such a process might take millions of years, producing a truly lasting legacy for an object that did so much for our understanding of the death of massive stars.

ABOVE: A string of cosmic pearls: the bright beads are where the shock wave from the explosion is hitting material ejected before the explosion.

ABOVE: Hubble's image of SN1987A in the heart of the Large Magellanic Cloud.

LEFT: The Large Magellanic Cloud before and after the supernova, which can be seen in the right image, just below the bright area located left and above centre, which is the Tarantula Nebula.

OPPOSITE: This wide-field view of the Small Magellanic Cloud shows many newly formed stars.

THE THIRD SISTER

Distance from Earth: 200,000 light-years

ABOVE: The Small Magellanic Cloud from the Spitzer Space Telescope.

In addition to the Large Magellanic Cloud, there have been interactions between the Milky Way and another lesser galaxy too, the Small Magellenic Cloud, on a similar orbit, but currently further from the Earth than the Large Magellanic Cloud. The two underwent a close pass two and a half billion years ago, and evidence of their interaction is preserved in the form of a faint stream of stars and gas that link the two dwarf galaxies. These cosmic exchanges of material are an early indicator, on this stage of our journey, of how important it is for galaxies to merge.

As we'd expect, within this neighbouring irregular dwarf galaxy there are stars and clusters of all kinds. The brightest star-forming region in the Small Magellanic Cloud is NGC 346, which spans a region of space 200 light-years across. NGC 346 is classified as an open cluster of stars, indicating that this group of stars all originated from the same collapsed cloud of matter. The new stars are being born in the associated nebula that contains this clutch of bright stars.

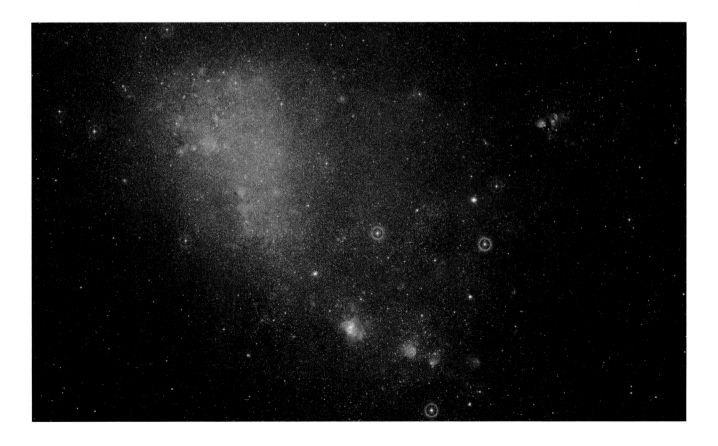

MEETING WITH A WANDERER

Distance from Earth: 300,000 light-years

We are at last on our way to the Great Spiral in Andromeda. We have a long way to go – 2.5 million light-years, but certainly we have left our own Galaxy behind, and we might expect to see nothing more, yet awhile. But then, unexpectedly, we encounter one system which we did not anticipate. It is a globular cluster, similar to those we have seen round the edge of our Galaxy, but well separated from it at a full 300,000 light-years away. This cluster really is on the galactic fringe.

It does not look bright from Earth simply because it is so far away, but now we are approaching it we can see that it is a perfectly ordinary globular cluster, neither particularly large or especially small; its diameter is approximately 520 light-years. It does not have an individual name of its own, but is known by its catalogue number of NGC 2419; and if we look around we will see very little, apart from the cluster itself. We are very much isolated in space, and it is not surprising that the cluster has been called an 'intergalactic tramp'. We now know that it is orbiting our Galaxy, but it takes something like three billion years to go round once.

Its extreme distance makes it difficult to study in detail from the Earth. It is unlikely, perhaps, to have formed in its current position, and may be the remnant of a galaxy that came too close and was captured by the Milky Way. There may even be other, smaller wanderers inhabiting these distant intergalactic regions, but there are very few of them and NGC 2419 serves as a reminder that, embedded within the Milky Way, we are in a privileged position – most of space is extremely empty indeed.

OPPOSITE: The Subaru Telescope's image of the globular cluster NGC 2419.

BELOW: Hubble Telescope image of NGC 2419.

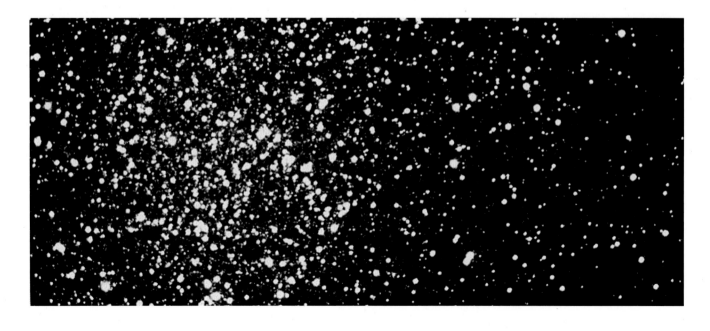

THE GREAT SPIRAL

Distance from Earth: 2,500,000 light-years

It used to be thought that the Andromeda Spiral, the nearest big galaxy to the Milky Way was both larger and more populous than the Milky Way system. It now seems that the two systems are about equal in mass, but M31 certainly contains more stars. The overall structure is the same as the Milky Way, with the visible galaxy formed of a flat disc with spiral arms, surrounded by an attendant retinue of globular clusters which trace the extent of Andromeda's dark matter halo. M31 will also have a massive black hole at its centre, but like the Milky Way's own black hole, it is relatively quiescent.

Although we are now more than two million light-years from Earth, our Galaxy is still close enough to reveal individual stars without too much trouble, including the all-important Cepheids that allow the cosmic distances to be measured. Andromeda is, in fact, approaching the Milky Way, and the two will likely undergo a dramatic collision in the small matter of a few billion years time. This collision, on a vastly grander scale than the interactions between the large and small Magellanic clouds and with the Milky Way, will result in the formation of a single massive galaxy, and in an enormous burst of star formation. Such collisions may have happened before, and the presence of what seems to be a double nucleus at the centre of Andromeda suggests that it may, itself, be the remnant of a major cosmic collision. There have been some suggestions that Andromeda has already interacted with the third of the Local Group's major galaxies, M33, and a lot of effort is being expended to map the space between them, with stars associated with Andromeda being uncovered tens of millions of light-years from the main galaxy.

Whatever the Andromeda Galaxy's past, it seems that it might be undergoing a dramatic change right now. As we have already mentioned, it has spiral arms, just like the Milky Way, but which are more tightly wound than ours, although just as distinct. However, a switch to infrared imaging shows that the predominant form of the galaxy is not spiral, but rather composed of a series of rings of gas and dust. One ring is particularly prominent, with a radius of about 30,000 light-years, which is about the same as the distance from the Sun to the centre of the Milky Way. If these structures represent the distribution of star formation today (which, given the presence of dust, seems likely), then the galaxy will slowly shift shape, as the bright blue stars that mark the spiral arms die, and new, ring-bound stars form.

The disc of Andromeda itself is warped, suggesting that recent disruption has taken place. If so, it supports the idea that this ring is probably the result of a recent interaction with one of Andromeda's many satellite galaxies. Some, such as the prominent M32, are large enough to be proper galaxies in their own right, but we must visit the third large member of the team before heading into deep space.

ABOVE: The Andromeda Galaxy can be seen with the unaided eye from dark locations. Here it is seen from La Silla in Chile, photographed by Serge Brunier.

RIGHT: The Andromeda Galaxy M31 imaged by Earth-based observer Greg Parker. The sister galaxy M32 is to the right of M31.

OUR GALACTIC GANG

Distance from Earth: 3,000,000 light-years

The Andromeda Galaxy and the Milky Way are not the only denizens of the local group. In addition to a generous handful of smaller galaxies, there is one more member that we should visit before leaving our local neighbourhood for deep space. This is the Triangulum Galaxy, M33, with only 40 billion stars or so, not so large as M31, or the Milky Way, but an equally beautiful spiral. It is, unlike Andromeda, tilted at a sensible angle for study from Earth. Just as with the Milky Way, there is a distinct nucleus at the centre of the spiral system, but there is no sign of a bar – and this absence seems to have a profound effect on a galaxy (as the bar can both trigger and impede the process of star formation), as well as channel material toward a system's central black hole.

M33's central black hole is rather puny, but it does host one record-breaker. M33 X-7, first detected via its X-ray emissions, is a 14 solar-mass black hole (14 times the mass of our Sun), the record-holder for a stellar-mass black hole (a category which excludes the behemoths lurking at the centre of galaxies which can be of the order of millions of solar masses).

ABOVE: The dwarf elliptical galaxy NGC 205, seen at bottom left of this picture, is another companion galaxy to M31, partially seen at top right.

BELOW: The Triangulum Galaxy M33, seen in infrared by the Spitzer Space Telescope.

HAVE A CIGAR

Distance from Earth: 11,500,000 light-years

ABOVE: This infrared image from the Spitzer Space Telescope captures extended dust features which do not show up in visible light.

BELOW: A Hubble Space Telescope image taken through four colour filters to capture visible and infrared light, as well as the light from the hydrogen filaments.

Some galaxies put on a show for our tour. One of them is our next stop, M82, a spectacular train wreck of a galaxy that is being consumed by a display of cosmic fireworks. It is, or at least was, a normal spiral galaxy, but now from the galaxy's centre a powerful wind is blowing, carrying gas at enormous speed, more than 125 miles per second (200 kilometres per second), up and away from the main disc.

We don't know what started this spectacular display, which seems to be powered by a massive bout of star formation, perhaps triggered by one or more close encounters with its neighbour, M81. Such a spectacular outburst produces stars of all sizes, including plenty of the massive stars which will quickly burn through all their fuel and explode in supernovae. Such supernovae may or may not be enough to power the dramatic wind, but are almost certainly powerful enough to set off further star formation in the disc around them, which is beautifully studded with bright star clusters. At least two hundred of these clusters exist, and the overall star formation rate is at least ten times that of the Milky Way. How unusual is M82? It's difficult to say, because we don't know how long the firework display is going to last. It may be that this is a one-off, a spectacular result of a particular set of circumstances. Or, it may be that most galaxies undergo spectacular episodes like this, before settling back down into a quiet life. Whichever it is, we're lucky to have a close-up view of such an event.

THE WHIRLPOOL GALAXY

Distance from Earth: 23,000,000 light-years

Further away we pause to visit a galaxy which is amazingly beautiful. Its catalogue number is M51, but for reasons that are obvious, even from a casual glance, it is always known as the Whirlpool. It is a 'grand design' spiral, connected to a smaller spiral by what looks like an extended arm, a trail of gas and stars. This smaller galaxy plunged through the disc of the larger galaxy around 500 million years ago, creating a shock that may have triggered the formation of the spiral arms themselves. A second pass, just 50 to 100 million years ago, may have caused further disruption.

Spiral arms in galaxies such as the Whirlpool are not physical features in the normal sense. They do rotate around the centre of the galaxy, but stars join and leave them as they travel on their own paths around the centre. The situation is remarkably similar to a traffic jam on a motorway – cars join at the back of the jam, and slowly work their way up to the front, where they can finally leave. The jam persists (and moves backwards along the motorway relative to the direction of travel of the cars), but it is always made up of different cars. Back in the Milky Way, then, the Sun must have passed through a spiral arm several times, and there are some (admittedly murky) hints in the fossil records that such times may have coincided with the periods of great extinctions (for example when the dinosaurs died out) on Earth. Such speculation isn't well supported, but at least the way that the arms move is well established.

As well as potentially causing the spiral arms, interactions with its neighbouring galaxy may have left the Whirlpool with a rather distinctive pattern of star formation. Unlike other spirals which leave star formation to the arms, rather than the normally quiescent bulge, there appears to be rapid star formation within the central 10,000 light-years of the Whirlpool. As much as four Sun's worth of stars may be being created every year, a rate that exceeds that achieved by the entire Milky Way. Perhaps this galaxy is in the process of forming its central bulge, or perhaps this rapid formation will have only a small effect on the overall structure of the galaxy. In either case, it will be interesting to see how the galaxy develops over the next few hundred million years, and tourists are well advised to return often.

ABOVE: The Whirlpool Galaxy NGC 5194 left and its companion NGC 5195, to the right, which is believed to be passing behind the Whirlpool, as seen from the Hubble Space Telescope in Earth orbit.

OPPOSITE TOP: The Whirlpool seen at visible wavelengths from the Hubble Space Telescope.

OPPOSITE BOTTOM: The Hubble Space Telescope's infrared view of the same galaxy, showing clumps of new-born stars that cannot be seen in visible light because of obscuring dust.

COLLIDING GALAXIES

Distance from Earth: 45-65,000,000 light-years

At the beginning of the intergalactic phase of our odyssey, we saw that the Andromeda Galaxy and the Milky Way were on a collision course. Such mergers are common; M33 might have collided with Andromeda in the past, and we've seen the wreckage of M82. Our latest stop – the Antennae Galaxies – are a spectacular glimpse into the future of our home Galaxy.

Part of a larger group with five other galaxies, the two galaxies are marked by long streams of stars stretching away from the colliding nuclear regions. The encounter between two previously separate galaxies began hundreds of millions of years ago, with the first real collision coming about one hundred million years ago. Remarkably, as with most galaxy mergers, there will probably have been no collisions between stars; the gaps between the stars are large enough that such encounters are rare. Gas clouds will collide, however, triggering dramatic bursts of star formation, visible today in the form of superclusters of hot, dense stars within the nuclei of the two galaxies. Most of these superclusters will not last long, with ninety per cent due to disperse in the next ten million years. In this way, the merger will have contributed to the smooth distribution of stars in the galaxy's final incarnation, which is likely to be a massive elliptical galaxy. The most massive clusters may become globular clusters, scattered throughout the new galaxy's halo, but really this is speculation.

Gravitational interaction between stars will also be common, with material being flung far from the centre of mass of the systems, creating those long, distinctive tails. Most of this material will remain gravitationally captured by the newly forming system, and it will eventually fall back into the galaxy. This sort of 'major merger' may be a frequent event for large galaxies, perhaps occurring as often as every few billion years. In any case, mergers make a major contribution to the build up of the largest galaxies in the present-day Universe. The ultimate fate of a merger can be determined with careful simulation, and depends on a whole host of parameters. The speed of approach, the angle of approach, the type of galaxies involved and their relative size all matter.

In the case of the Antennae, we are probably looking at a collision between two spiral galaxies, which provided the fuel for the spectacular events that followed. Not all mergers will trigger star formation like this one, though; collisions between more evolved galaxies (such as two ellipticals) will act as a 'dry merger', combining two old, red and dead galaxies to form nothing more than a larger, old, red and dead galaxy.

ABOVE: This Earth-based image shows the incredible merging tails of the Antennae Galaxies.

ABOVE: Composite image from the Hubble's images in visible and infrared with views at millimetre wavelengths from ALMA, the Atacama Large Millimetre/sub-millimetre Array of radio telescopes, in the Atacama desert of northern Chile.

A CITY OF STARS

Distance from Earth: 53,000,000 light-years

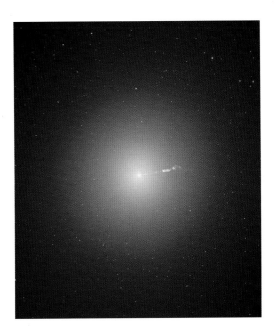

ABOVE: A remarkable feature of the giant elliptical galaxy M87 is a jet of electrons and other subatomic particles travelling at near the speed of light, powered by the black hole at the centre of the galaxy.

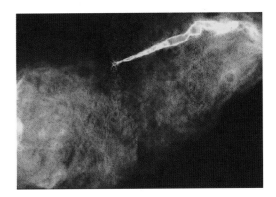

ABOVE: M87's jet seen at radio wavelengths by the Very Large Array of radio telescopes.

So far we have visited several small groups of galaxies, each containing a few large systems and their attendant retinues. Larger groups exist, and we are now flying through the outskirts of one of the largest of these galaxy clusters, the Virgo cluster. As the galaxies become more tightly packed, mergers become more common and the systems we're passing are mostly red, dead ellipticals, their reserves of gas for star formation exhausted by repeated collisions. The largest of them is M87, the monster at the heart of the Virgo Cluster's labyrinth.

Everything about M87 is spectacular, starting with the globular clusters that we encounter on our approach. The galaxy has more than 12,000 of them, compared to just a few hundred around the Milky Way. There may be as many stars in these clusters alone as there are in our entire galaxy, a humbling thought for a traveller from a more modest home coming across M87 for the first time.

The galaxy itself is 200 times more massive than the Milky Way, and may contain as many as a trillion (a thousand billion) stars. They form an almost featureless ball, unmarred by the dust lanes that streak lesser galaxies. M87 has grown so large over the eons that even the largest of its companion galaxies can be swallowed without even rippling its surface. Even without such mergers, it is steadily growing ever larger, absorbing a few Sun's-worth of matter from the cluster each year.

Most of this material will be accreted onto the central region of the galaxy, a black hole that it is almost certainly a few billion times the mass of the Sun. There is a close link between the size of a galaxy, and the size of its black hole, so there is no surprise in finding such a monster at M87's heart, but it is still impressive. Material, even here, doesn't accrete directly onto the black hole, but forms an immense accretion disc, nearly half a light-year across, feeding the black hole the equivalent of a Sun's-worth of mass every ten years.

This level of activity might be normal for M87, but it is spectacular by any standards, and it is no surprise to find that it is powering an enormous jet. Material is expelled from the inner regions of the accretion disc travelling at a substantial fraction of the speed of light. It is possible that such jets had a critical role to play in the evolution of M87 and its fellow ellipticals, but the galaxy is now large enough to make its presence nothing more than a curiosity. It is certainly spectacular, and can even be seen visually in the largest amateur telescopes back on Earth.

THE VIRGO CLUSTER

Distance from Earth: 53,800,000 light-years

As we travel amongst the galaxies, it's easy to lose track of where we are. Each galaxy looks different – its personality and its history revealed by its shape, whether it is a beautiful barred spiral, or one of the more simple types of galaxy such as an elliptical. The Virgo elliptical galaxies, which include some of the largest in the present-day Universe, are essentially just balls of stars, each of which is still orbiting around a central black hole, but in which all traces of a coherent disc are absent.

The typical elliptical galaxy is different from its spiral cousins in many other ways, too. A quick look at the galaxies around us, as we enter the Virgo cluster, shows that the blue stars that mark recent and vigorous star formation are absent from the vast majority of ellipticals. The cluster is the largest nearby concentration of galaxies, home to thousands of large systems, compared to the two or three in the local group, which is being pulled towards it. As we get deeper into the cluster itself, we begin to lose sight of the few familiar spiral galaxies altogether; in dense environments it is the ellipticals which dominate.

This probably suggests that mergers between galaxies, which as we have seen can rapidly use up the fuel for star formation, are common here; and that's true, at least, on the cluster's outskirts. As we get further in, the galaxies are caught in the grip of the cluster's immense gravitational pull, and are moving too quickly to inflict more than a glancing blow on their neighbours.

To look just at the galaxies, though, is missing the point somewhat. Less than half of the mass of the Virgo cluster lies in the galaxies, with the rest being primarily hot gas that fills the space between them. Some of this gas may be primordial, dating back to the time when the cluster was forming, but the rest may have been expelled from galaxies during mergers on the cluster's edge. The gas is not static, either, with immense flows of material transferring energy from the cluster's dense heart to the edge.

Amongst the thousands of ellipticals that make up the Virgo cluster, it is hard to distinguish one from another. Lurking at the centre, though, is the largest galaxy we have encountered so far, and it is not quiet.

OPPOSITE: There are over one thousand galaxies in the Virgo cluster, the closest large cluster of galaxies to our own Local Group. In the central portion of this image, faint foreground dust hangs above the plane of our own Milky Way, and the dominant member of the cluster, M87, is just below the centre of the frame.

STRANGE GREEN OBJECT

Distance from Earth: 650,000,000 light-years

This mysterious green object, tens of thousands of light-years across – a decent size for a galaxy – is unique in the exploration of the Universe. The colour green indicates glowing oxygen, and it appears to have a vast central hole, some 16,000 light-years across.

It gets its name from a Dutch school teacher – Hanny van Arkel – who discovered it while participating in the GalaxyZoo.org project – an astronomy research project started by one of us (CL), involving a quarter of a million volunteers, classifying hundreds of thousands of galaxies. The project uses data from the Sloan Digital Sky Survey and more recently from the Hubble Space Telescope. Voorwerp, it turns out, just means 'object' in Dutch!

There are various hypotheses about the nature of the object, but it seems it is being stimulated to emit its light by activity in an adjacent galaxy (IC 2497). The Voorwerp shows clear signs of being disrupted by a recent merger, although there is no sign of a partner galaxy with which it might have collided. The merger will have funnelled material toward the black hole in the centre of IC 2497, which will have undergone an intense period of activity, bright enough to be considered a quasar – a term for the core of a galaxy in which a massive black hole is consuming huge amounts of gas and dust, and radiating exceptionally brightly. As we saw at M87, such activity is often associated with the production of jets, a galaxy-scale equivalent of what we saw in the case of the cosmic corkscrew, SS433.

The Voorwerp is located in exactly the right position to be hit by just such a jet, and radio observations have shown what seems to be the remnants of one. However, there is a mystery. If there was once a quasar in IC 2497, it is not there today. Any such object would shine brightly in X-rays, and observations with space-based X-ray telescopes have drawn a blank. It must have subsequently switched off, making the Voorwerp the only place where we can study what happens when a galaxy switches from active to quiescent. Such shifts may, it seems, happen relatively frequently, and the population of quasars suddenly changes over time. Relatively dominant a couple of billion of years ago, quasars are much less common today.

Not all the Voorwerp's secrets have been revealed. There appear to be stars forming in the part of the cloud closest to IC 2497, perhaps induced by the impact of the jet on the Voorwerp's gas. The circular hole we noticed at the start of our visit is also still unexplained. Perhaps it is the shadow cast by material close to the former quasar, but this explanation seems unconvincing. Further work is needed!

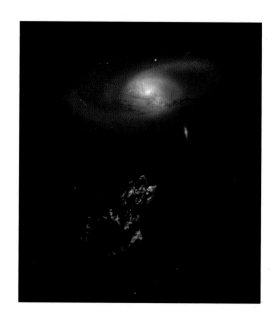

ABOVE: The Voorwerp's green light, which is produced by ionized oxygen atoms, is suspected to be part of a tidal tail of material illuminated by a quasar inhabiting the centre of the adjacent galaxy IC 2497.

BENDING LIGHT

Distance from Earth: 8,000,000,000 light-years

ABOVE: This close-up of the Abel cluster of galaxies shows gravitational lensing in action. Light from all the galaxies behind the cluster is stretched into red, orange and blue arcs, allowing astronomers to see faint objects that could not otherwise be seen.

We have travelled a long way out into the Universe, but despite the wonders all around us it is still tempting to look back towards home. The Sun is now impossible to pick out, even with our remarkably advanced instruments on the Ptolemy, but the Milky Way is still there, visible as part of the distinctive local group of three spirals.

As we've moved further away, something strange has begun to happen. The image of our distant home has become distorted, and the apparent shape of the Milky Way has transformed into an arc. The effect is strongest when we are peering through a galaxy cluster, and from this far out it is almost impossible to find an undistorted view.

The effect we are witnessing is known as gravitational lensing, the bending of light by the presence of mass. Such an effect has been known about for a long time; just as we see these distant galaxies distorted by the journey of their light through the Cosmos, scientists in the early twentieth century predicted that the apparent positions of the stars would shift if they could be seen when the Sun passed close-by.

An effect of this nature is predicted by Newton's theory of gravity, although he'd be surprised to see it on these scales, but when the size of the shift was measured during a series of eclipse expeditions (the eclipse being necessary to make viewing of stars near the Sun possible) the results agreed with the predictions of Einstein's theory of relativity, a critical proof of the new theory's superiority.

Since then, astronomers have learnt to make use of gravitational lensing in all sorts of ways. By measuring the amount of distortion induced in passing light they are able to weigh galaxy clusters, and by measuring subtle lensing effects they can map the distribution of mass in the Universe. If everything is lined up correctly, the lensing effect can function as a natural telescope, magnifying the most distant galaxies and making them easier to see. In our case, the Milky Way may appear distorted from its familiar shape, but thanks to lensing it shines just a little brighter all the way out here, billions of light-years from home.

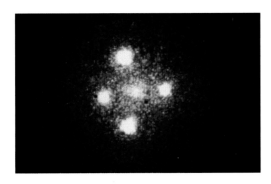

ABOVE AND RIGHT: The Einstein Cross, or Q2237+030 or QSO 2237+0305 is a gravitationally lensed quasar that sits directly behind ZW 2237+030 (Huchra's Lens). Four images of the same distant quasar appear around a foreground galaxy, due to strong gravitational lensing.

TO INFINITY AND BEYOND

Distance from Earth: 13,000,000,000 light-years

Our journey has been limited only by our imagination and by the information that we can gather about the Universe, in order to let us plot our route amongst the stars and galaxies. Our tour is coming to a close, but how far could we travel? Some of the most distant galaxies we know about have been seen in incredible deep fields, the result of pointing a large telescope at an apparently empty and boring patch of sky for nights on end, allowing images of the faintest of galaxies to build up over time.

The most famous of these deep fields comes from the Hubble Space Telescope, which was the first to carry out this sort of survey in a serious way. Most of the galaxies in this image are seen from Earth as they were in the Universe's first few billion years. As we might expect, the deep fields contain young galaxies which only have stars, with enormous reservoirs of gas still available for star formation. With relatively little time for mergers to take place, they still retain a host of irregular and bizarre shapes rather than having settled down to the familiar spirals and ellipticals we see today.

Wouldn't it be wonderful to see this early Universe first-hand? Unfortunately, just because Ptolemy can travel faster than light, it doesn't mean we can travel backwards in time, and arriving in Ptolemy at the part of space that contains the most distant of the Hubble Deep Field galaxies is a little disappointing, since it looks much like the region around our Milky Way. If we looked carefully enough, we could probably find a group of three spiral galaxies, two of which have interacted in the past and two of which are on a collision course, a close analogue to our own Local Group. Although telescopes back on Earth see this region as it was when the light now arriving left, more than ten billion years ago, of course time has passed here just as it has at home, and we cannot learn any more about the early days of the Universe by travelling to places even billions of light-years away.

In some ways it is reassuring that one region of space looks much like another. A fundamental assumption of modern cosmology maintains that there is nothing special about our tiny part of the Universe, and therefore we are justified in drawing conclusions about the whole from the little we can see. Indeed, if our present ideas about how the Universe began are right, a rapid expansion in the first fraction of a second after the Big Bang must mean that the entire observable Universe is but a tiny fraction of the vast grandeur of space.

ABOVE: This Hubble Ultra Deep Field view of the most remote galaxies looks back in time to 400–800,000 years after the Big Bang. The whole image (this is just a segment) contains an estimated 10,000 galaxies.

ECHOES OF THE BIG BANG

Distance from Earth: 13,700,000,000 light-years

We may not be able to travel to the early Universe directly, but throughout our journey radiation (left over from those earliest days of the Universe), has been our constant companion. This Cosmic Microwave Background radiation is light that was last scattered something like 400,000 years after the Universe came into existence – as it cooled, after beginning its life in the hot, dense state we call the Big Bang. Soon after the time of its last scattering, the radiation would have been relatively bright in the visible and infrared, but the expansion of the Universe over a long passage of time has cooled it, and stretched out its wavelength to peak in the microwave region of the spectrum today. This peak wavelength corresponds to a temperature of approximately –270.3 degrees Celsius, which at these very low temperatures is usually expressed as 2.7 degrees Kelvin, just a few degrees above absolute zero (absolute zero is defined as 0 degrees Kelvin, –273 degrees Celsius, and is the lowest possible temperature, where nothing could be colder and no heat energy remains in a substance).

Of course, one doesn't have to travel as far as we have to see the microwave background radiation. It can be detected and studied from the ground – where it accounts for a few per cent of the 'white noise' picked up by older, analogue, radio or television receivers – and from space, through satellites like the American WMAP (Wilkinson Microwave Anisotropy Probe) and the European Planck spacecraft. The tiny fluctuations in temperature observed in this background radiation give us our earliest glimpse of the Universe, and it is from these tiny, apparently insignificant, seeds, that everything we have encountered on our grand journey of the imagination, from planets to stars to galaxies and even the majestic galaxy clusters themselves, has grown. Much has happened in the 13.7 billion years since this primordial radiation first filled the Cosmos, but perhaps the most remarkable event of all is the evolution of a species that, sitting on a perfectly normal planet orbiting an ordinary star, can embark on an imaginary tour of the Cosmos!

BELOW: Colours in the WMAP image represent temperature fluctuations of the remnant glow from the early Universe: red regions are warmer and blue are colder. The map was compiled from five years of collected data.

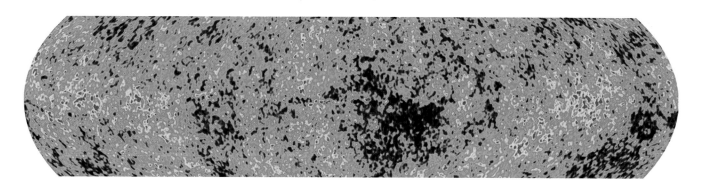

EPILOGUE

Now we are back in our armchairs, with all those great memories of a tourist trip of a lifetime, there is much to ponder.

At the end of the outward journey, just before we turned the Ptolemy's drive back towards home, we were at the very edge of what we call the 'Observable Universe' – that part of the whole that can be seen from Earth.

We were cruising around among objects like the ones in the Hubble Ultra Deep Field – all those oddly-shaped young galaxies that indicate, as viewed from Earth, that we are seeing the state of the Universe only a short time after its birth. But while we were actually out there, sitting in the Ptolemy, we saw them not at all like this, the way they looked near the dawn of creation. Having got there at the speed of thought, we saw those galaxies as they are now, the survivors of a long history of evolution and collisions with other galaxies, looking just like the ones that are close to our own Milky Way. Indeed there seems no reason to believe that this undiscovered new territory would be any different from our local observable Universe, and cosmologists have enshrined this belief deep within their current theories; they call it the 'Cosmological Principle'.

But there is more. From our vantage point billions of light-years from Earth, we would have been able to see a whole new vista, a part of the Universe whose light has yet to reach those stuck back on Earth. We hadn't actually travelled to the edge of a Universe at all; we were still somewhere in the middle of one, with stars, galaxies, space and stuff in every direction, just we as we see space all around us from Earth. There, on the very edge of the Observable Universe, we could see just as much space ahead of us as we had left behind.

Because we travelled so far, we effectively doubled the size of the Universe we know about.

Now suppose we keep on going. How far could we get? How much larger is the Universe than the bit we can see from Earth? There are indications that the true answer is 'very much larger', and the Universe may even be infinite. If the Universe truly is infinite in size, then - even without any hypothesising about parallel Universes - somewhere out there in the oceans of space will be another you, another us, another Cosmic Tourist, each exactly like our own.

That's what infinity means: no matter how small the chance of such a coincidence, given an infinitely large Universe, somewhere, the unlikely event will happen.

So let us rejoice! We are not alone on our cosmic journey, even though, sadly, we will never meet our identical twins.

And of course, in an infinite Universe, we should never say never.

PRACTICAL ADVICE FOR ARMCHAIR TOURISTS

Our tour of the Cosmos has been very much one of the imagination; the distances are impossible to conceive. What motivated the journey in the first place was curiosity about what is 'out there' and whether we can make any sense at all of our humble situation on planet Earth in a Universe that is 'infinitely' large. Cosmologists have different views on whether the Universe is technically of infinite size, but from a human perspective it may as well be, since even taking the first step out to the nearest star would take many lifetimes at the speeds we are currently able to achieve. But what we can do is to explore it with our eyes, assisted with binoculars or telescopes. In this brief guide the numbers shown (1–100) refer to the numbers we have allocated to each stop on the tour.

When it comes to defining the brightness (the so-called 'magnitude') of heavenly bodies that can be seen with the naked eye, binoculars or telescopes, observers differ in the limits they can see, so guidance on what you will be able to see is not exact. The scale used for magnitude dates back to the time of the ancient Greeks who divided the stars that are visible to the naked eye into six classes: the brightest were magnitude 1, and the faintest magnitude 6. Each grade of brightness was considered to be twice the brightness of the following grade (this is really the basis of a 'logarithmic scale'). The modern magnitude scale still works in the same way as a golfer's handicap, with the more brilliant performers having the lowest values, but it has been extended to encompass the very brightest, which is our Sun with a magnitude of –26.74, and the very dimmest, which are the magnitude 30 stars observed by the Hubble Space Telescope.

The Naked Eye

It is surprising how much can be seen with the naked eye alone! During the daytime the sky is of course very bright and the Sun (12–15) is dominant, but there are times when the only way to observe a bright comet is when the Sun is above the horizon, though this is very rare. Looking straight at the Sun with the naked eye is definitely forbidden as you will risk blindness. To view sunspots on the face of the Sun, or eclipses, the Sun must be projected onto a suitable background using a telescope. You can then follow a spot as it tracks across the disc of the Sun, or observe the progress of the Moon across the face of the Sun in an eclipse.

Sometimes the Moon (4–11) can also be observed in the daytime; Venus (20–21), the brightest planet, can also be seen in daylight if you know where to look.

At night-time, the power of the naked eye becomes very evident. The Moon can be located without difficulty, and with the naked eye the maria, the great dark plains which cover so much of the lunar surface, are easy to see. The most prominent are Mare Imbrium (10), and Mare Serenitatis (4), and Tranquillitatis (4) and some of the larger craters such as Copernicus (8), Aristarchus (9) and Plato (10) are easy to see. It is also possible to see the Earthshine – that is to say, the night-side of the Moon, shining by the light reflected from the Earth.

With the naked eye, the bright planets, Venus (20), Jupiter (37), Saturn (43) and Mars (26), can be seen, but only Venus and Jupiter are really obvious. Stars can be seen down to magnitude 6. Sirius (56), the brightest star, has a negative value of –1.6; the planets greatly exceed this. Venus has a magnitude of above –4. It is interesting to see what your personal limits are. Some people can see stars down to magnitude 7. Fomalhaut (59), the star with an extrasolar planet can also been seen very easily – it has a bright magnitude 1.73, though of course the Hubble Space Telescope is needed to identify the planet! Castor (60) and its companion Pollux in Gemini, the Twins, are easy stars to find; Algol (61) at magnitude –0.15 is an even brighter naked-eye star.

Bright open clusters such as the Pleiades (64) are easy to find. Nebulae are in general below the limit for the naked eye. The globular cluster Omega Centauri (81) is easy to see with the naked eye in the southern hemisphere, where the Large (86) and Small (88) Magellanic Clouds are obvious. The band of the Milky Way (84–85) is one of the finest sights in the sky.

The Zodiacal Light (22) may be seen as a cone of light after sunset – very difficult from countries with light pollution but with clearer skies it can be quite conspicuous, as bright as the Milky Way.

ABOVE: 10 x 50 binoculars.

Binoculars

Choice of binoculars depends upon what you want to see. I have a pair of 7 x 50s (regular binoculars) which give splendid views over a wide area. Open clusters like the Pleiades (64), the Milky Way's star fields (84–85), and the moons of Jupiter (37) are evident, and the one bright satellite of Saturn – Titan (45). Saturn itself (43) shows up as a curious shape. To see Saturn with its rings you really need a telescope.

A mass of detail is evident on the Moon with 7 x 50s – everything we have visited in Ptolemy apart from the Far Side of course. Some nebulae can be seen with binoculars – I cannot see the Ring Nebula (74) with the low-power binoculars, but can with the high-power (20 x 80) binoculars. The Andromeda Galaxy (90) looks like a smudge in binoculars! The Large and Small Magellanic Clouds ((86–88) appear like broken off parts of the Milky Way, which is effectively what they are. Quite a number of asteroids (33–34) can be identified if you know their precise positions – they look exactly like stars and betray their real nature only because they move from night to night-time since they are Solar System bodies.

Comets bright enough to be seen with binoculars are quite common and look rather like out of focus stars. Some amateurs use high-powered binoculars to sweep the skies for new comets; British amateur observer George Alcock discovered eight comets using his 20 x 80 binoculars.

Telescopes

Use a small telescope to project the Sun to see sunspots and eclipses. A great amount of detail can be seen on the Moon – everything we saw from Ptolemy, apart from the Far Side of course, and the remnants of Apollo's visits.

ABOVE: 20 x 80 binoculars.

ABOVE: A typical modern 90-millimetre starter telescope with computer control.

ABOVE: Patrick Moore with his 15-inch reflecting telescope.

Viewing Mercury (17), the phase can be followed with a small telescope but no detail can be seen. For Venus (20), the phase is evident, and vague markings on the disc (the face of the planet) might be seen, as well as the Ashen Light (20). On Mars (26) you will be able to see its main features such as the dark area Syrtis Major (27), and the polar caps (26) waxing and waning as the ice forms and sublimes. Among the asteroids, Vesta (33), Pallas, Juno, Hygeia and perhaps a dozen others can be seen. They look like stars and, as before, only their movement across the night sky on consecutive nights betrays the fact they are inhabitants of the Solar System.

Looking at Jupiter (37), the belts will be clear, and you can see the Great Red Spot (38) and various small spots that appear periodically but do not persist for long. You can also see Jupiter's four main satellites (39–40), but they may be occulted (hidden) by Jupiter, or eclipsed by Jupiter's shadow.

Saturn (43) is always the most impressive sight in the heavens: the rings can be seen clearly, and under excellent conditions the Cassini Division (41) can be made out when the rings are wide open (when we see them from above so they appear wide). Titan (45) is easy to see, and Rhea (41) and Iapetus (44) are too; Tethys and Dione (45) can also be seen. With his three-inch telescope PM has even glimpsed Mimas (41–46) and Enceladus (46), but not easily.

Uranus (47) and its four main satellites are visible, and the green colour of Uranus's disc is very evident. On Neptune (48), the bluish colour can be discerned with a three-inch telescope and Triton (49) is easy to see. Of the Kuiper Belt objects only Pluto (50) is within range of a small telescope – just!

Stars: the colours of the stars are very evident in a small telescope, particularly red stars like Betelgeux (67), which is an exceptionally lovely sight. Plenty of double stars are available, some, such as the southern Alpha Centauri (55), are striking. Any good star catalogue will list many other stars which can be separated with a small telescope. With a three-inch telescope, generally speaking any double star can be seen as such if both components are brighter than the 11th magnitude, and sufficiently widely separated.

Variable stars: plenty of those are within reach; cepheids for example, whose changes can be followed from night to night. Mira (63), the red long-period variable, is outstanding and the brightest star of its type (also a naked-eye object for several weeks every year). With a telescope, Mira can be followed for months; at minimum it drops out of range for a small telescope.

Catalogues will list long period variables which rise above the tenth magnitude at maximum. Some variables have a surprisingly large range. Chi Cygni can reach the third magnitude on occasion and falls to about 14 at minimum; so it then becomes a difficult object even for larger telescopes.

Star clusters: plenty of these can be found. The open clusters can be condensed or sparse. The Pleiades (64), Hyades and Praesepe are the best-known examples. The globular cluster Omega Centauri (81) will show a great number of stars, and NGC 2419 (89) will also be visible.

Galaxies: these are visible if they have a combined magnitude above about 11, so you should be able to see the Cigar (92), the Whirlpool (93), and M87 (95), though the Antennae (94) and the Virgo Cluster (96) may prove too challenging with this size of telescope.

GLOSSARY

ABERRATION of starlight. As light does not move infinitely fast, but at a rate of 186,000 miles per second (300,000 kilometres per second), and as the Earth is moving round the Sun at an average velocity of 15 miles per second (25 kilometres per second), the stars appear to be shifted slightly from their true positions.

ABLATION. The erosion of a surface by friction or vaporization.

ABSOLUTE magnitude. The apparent magnitude that a star would have if it could be observed from a standard distance of 32.6 light-years (1 parsec).

ABSOLUTE zero. The coldest possible temperature: −273.16 °C.

ACCRETION disc. A disc structure which forms round a spinning object when material falls on to it from beyond.

AIRGLOW. The light produced and emitted by the Earth's atmosphere (excluding meteor trails, thermal radiation, lightning and aurorae).

ALBEDO. The reflecting power of a planet or other non-luminous body. The Moon is a poor reflector; its albedo is 7% on average.

ALTITUDE. The angular distance of a celestial body above the horizon.

ÅNGSTRÖM unit. One hundred-millionth part of a centimetre.

APHELION. The furthest distance of a planet or other body from the Sun in its orbit.

APOGEE. The furthest point of the Moon from the Earth in its orbit.

APPARENT magnitude. The apparent brightness of a celestial body. The lower the magnitude, the brighter the object: the Sun is approximately −27, the Pole Star +2, and the faintest stars detect able by modern techniques around +30.

ARRAY. An arrangement of a number of linked radio antennae.

ASTERISM. A pattern of stars which does not rank as a separate constellation.

ASTEROIDS. A class of small solar system bodies in orbit around the Sun.

ASTRONOMICAL unit. A.U. The mean distance between the Earth and the Sun. It is equal to 92,955,887.6 miles (149,598,500 kilometres).

AURORA. Aurorae are 'polar lights'; Aurora Borealis (northern) and Aurora Australis (southern). They occur in the Earth's upper atmosphere, and are caused by charged particles emitted by the Sun.

AZIMUTH. The bearing of an object in the sky, measured from north (0°) through east, south and west.

BAILLY'S beads. Brilliant points seen along the edge of the Moon just before and just after a total solar eclipse. They are caused by the sunlight shining through valleys at the Moon's limb.

BARYCENTRE. The centre of gravity of the Earth Moon system. Because the Earth is 81 times as massive as the Moon, the barycentre lies well inside the Earth's globe.

BASALT. A dark grey fine-grained volcanic rock, with a silica content of from 44 and 50%; basalt is the most widespread volcanic rock found on the surfaces of the terrestrial planets.

BILLION. (American) One thousand million. (British) One million million. The American version is now generally used.

BINARY star. A stellar system made up of two stars, genuinely associated, and moving round their common centre of gravity. The revolution periods range from millions of years for very widely separated visual pairs down to less than half an hour for pairs in which the components are almost in contact with each other. With very close pairs, the components cannot be seen separately, but may be detected by spectroscopic methods.

BLACK body. A body which absorbs all the radiation falling on it.

BLACK hole. A region round a very small, very massive collapsed star from which not even light can escape.

BODE'S law. A mathematical relationship linking the distances of the planets from the Sun. It may or may not be genuinely significant.

BOW shock. The edge of the magnetosphere of a planetary body, where the solar wind is deflected.

BROWN dwarf. Faint stars of 0.01-0.08 solar mass whose core temperatures are not high enough for nuclear reactions to take place.

CALDERA (calderae). A large depression, usually found at the summit of a shield volcano, due to the withdrawal of magma from below.

CARBON stars. Red stars of spectral types R and N with unusually carbon-rich atmospheres.

CARBONACEOUS chondrites. Primitive stony meteorites (aerclites), containing carbonaceous compounds and hydrated silicates.

CELESTIAL sphere. An imaginary sphere surrounding the Earth, whose centre is the same as that of the Earth's globe.

CEPHEID. A short-period variable star, very regular in behaviour; the name comes from the prototype star, Delta Cephei. Cepheids are astronomically important because there is a definite law linking their variation periods with their real luminosities, so that their distances may be obtained by observation.

CHONDRITE. Stony asteorite containing chondrules. Chondrites make up over 90% of stony meteorites (aerolites).

CHONDRULES. Spherical incursions found in chondrites. They are composed mainly of pyroxene and olivine, with some glass.

CHROMOSPHERE. That part of the Sun's atmosphere which lies above the bright surface or photosphere.

CIRCUMPOLAR star. A star which never sets (dips below the horizon). For instance, Ursa Major (the Great Bear) is circumpolar as seen from England; Crux Australis (the Southern Cross) is circumpolar as seen from New Zealand.

COLLAPSAR. The end product of a very massive star, which has collapsed and has surrounded itself with a black hole.

COLOUR index. The difference between a star's visual magnitude and its photographic magnitude. The redder the star, the greater the positive value of the colour index; bluish stars have negative colour indices. For stars of type A0, colour index is zero.

CONJUNCTION. (1) A planet is said to be in conjunction with a star, or with another planet, when the two bodies are apparently close together in the sky. (2) For the inferior planets, Mercury and Venus, inferior conjunction occurs when the planet is approximately between the Earth and the Sun; superior conjunction, when the planet is on the far side of the Sun and the three bodies are again lined up. Planets beyond the Earth's orbit can never come to inferior conjunction.

CORONA. The outermost part of the Sun's atmosphere, made up of very tenuous gas. It is visible with the naked eye only during a total solar eclipse.

CORONAGRAPH. A device used for studying the inner corona at times of non-eclipse.

COSMIC rays. High-velocity particles reaching the Earth from outer space. The heavier cosmic-ray particles are broken up when they enter the upper atmosphere.

COSMIC year. The time taken for the Sun to complete one revolution round the centre of the Galaxy: about 225,000,000 years.

COSMOLOGY. The study of the universe considered as a whole.

COUNTERGLOW. The English name for the sky-glow more generally called by its German name of the Gegenschein.

CULMINATION. The maximum altitude of a celestial body above the horizon.

DAY, sidereal. The time taken for the Earth to rotate 360 degrees on its axis with respect to the background stars.

DAY, solar. The mean interval between successive meridian passages of the Sun: 24 h 3 m 56 s.555 of mean sidereal time. It is longer than the sidereal day because the Sun seems to move eastward against the stars at an average rate of approximately one degree per day.

DECLINATION. The angular distance of a celestial body north or south of the celestial equator. It corresponds to latitude on the Earth.

DOPPLER effect. The apparent change in wavelength of the light from a luminous body which is in motion relative to the observer. With an approaching object, the wavelength is apparently shortened, and the spectral lines are shifted to the blue end of the spectral band; with a receding body there is a red shift, since the wavelength is apparently lengthened.

DOUBLE star. A star made up of two components – either genuinely associated (binary systems) or merely lined up by chance (optical pairs).

DWARF novae. A term sometimes applied to the U Geminorum (or SS Cygni) variable stars.

DWARF star. In general, a small star in the hydrogen-burning phase of its evolution, therefore on the Main Sequence.

EARTHSHINE. The faint luminosity on the night side of the Moon, frequently seen when the Moon is in its crescent phase. It is due to light reflected onto the Moon from the Earth.

ECLIPSE, lunar. The passage of the Moon through the shadow cast by the Earth. Lunar eclipses may be either total or partial. At some eclipses, totality may last for approximately 1¾ hours, though most are shorter.

ECLIPSE, solar. The blotting-out of the Sun by the Moon, so that the Moon is then directly between the Earth and the Sun. Total eclipses can last for over 7 minutes under exceptionally favourable circumstances. In a partial eclipse, the Sun is incompletely covered. In an annular eclipse, exact alignment occurs when the Moon is in the far part of its orbit, and so appears smaller than the Sun; a ring of sunlight is left showing round the dark body of the Moon. Strictly speaking, a solar 'eclipse' is the occultation of the Sun by the Moon.

ECLIPSING variable (or Eclipsing Binary). A binary star in which one component is regularly occulted by the other, so that the total light which we receive from the system is reduced. The prototype eclipsing variable is Algol (Beta Persei).

ECLIPTIC. The apparent yearly path of the Sun among the stars. It is more accurately defined as the projection of the Earth's orbit on to the celestial sphere.

ELECTROMAGNETIC Spectrum. The full range of electromagnetic radiation from shortest to longest wavelengths. Gamma rays have the shortest wavelengths, less than 0.01 nanometres; X-rays have wavelengths from 0.01–10 nanometres; ultraviolet is 10–390 nanometres; visible light has wavelengths 390-700 nanometres; infrared is 700 nanometres to 1 millimetre; the longest wavelengths are radio from 1 millimetre extending upwards – for example long wavelengths are measured in kilometres.

ELECTROMAGNETIC radiation. This is the name for periodically changing electric and magnetic disturbance which travels in a vacuum at the speed of light.

ELONGATION. The angular distance of a planet from the Sun, or of a satellite from its primary planet.

EQUATOR, celestial. The projection of the Earth's equator onto the celestial sphere.

EQUINOX. The equinoxes are the two points at which the ecliptic cuts the celestial equator. The vernal equinox or First Point of Aries now lies in the constellation of Pisces; the Sun crosses it about 21 March each year. The autumnal equinox is known as the First Point of Libra; the Sun reaches it about 22 September yearly.

ESCAPE velocity. The minimum velocity which an object must have in order to escape from the surface of a planet, or other celestial body, without being given any extra impetus.

EVENT horizon. The 'boundary' of a black hole. No light can escape from inside the event horizon.

EXOLANET or Extrasolar Planet. A planet outside the Solar System.

EXTINCTION. The apparent reduction in brightness of a star or planet when low down in the sky, so that more of its light is absorbed by the Earth's atmosphere. With a star 1° above the horizon, extinction amounts to 3 magnitudes.

FACULAE. Bright, temporary patches on the surface of the Sun.

FIELD star. A star which is seen close to a stellar cluster, but is not a cluster member. It may be much closer or much more remote.

FLARES, solar. Brilliant eruptions in the outer part of the Sun's atmosphere. Normally they can be detected only by spectroscopic means (or the equivalent), though a few have been seen in integrated light. They are made up of hydrogen, and emit charged particles which may later reach the Earth, producing magnetic storms and displays of aurorae. Flares are generally, though not always, associated with sunspot groups.

FLARE stars. Faint red dwarf stars which show sudden, short-lived increases in brilliancy, due to intense flares above their surfaces.

FLASH spectrum. The sudden change-over from dark to bright lines in the Sun's spectrum, just before the onset of totality in a solar eclipse. The phenomenon is due to the fact that at this time the Moon has covered up the bright surface of the Sun, so that the chromosphere is shining 'on its own'.

FLOCCULI. Patches of the Sun's surface, observable with spectroscopic equipment. They are of two main kinds: bright (calcium) and dark (hydrogen).

FRAUNHOFER lines. The dark absorption lines in the spectrum of the Sun or any other star.

GALAXIES. Systems made up of stars, nebulae, and interstellar matter. Many, though by no means all, are spiral in form.

GALAXY, the. The system of which our Sun is a member. It contains approximately 100,000 million stars, and is a rather loose spiral.

GEGENSCHEIN. A faint sky-glow, opposite to the Sun and very difficult to observe. It is due to thinly spread interplanetary material.

GEODESY. The study of the shape, size, mass and other characteristics of the Earth.

GIBBOUS phase. The phase of the Moon or planet when between half and full.

GEOSYNCHRONOUS orbit. An orbit round the Earth at an altitude of 22,236 miles (35,786 kilometres), where the period will be the same as the Earth's sidereal rotation period (23h 56m 4.1s) assuming that the orbit is circular and lies in the plane of the Earth's equator.

GLOBULES. Small dark patches inside gaseous nebulæ. They are probably embryo stars.

GOLDILOCKS zone. The region around a star where a planet can maintain liquid water on its surface. So-called because conditions are neither too hot nor too cold, but just right for life to exist.

GRAVITATIONAL lensing. The bending of light from a distant object by a massive object lying between the object and the observer. This often results in the observer seeing several images of the distant object, and is one of the predictions of Einstein's General Theory of Relativity.

GREAT Bombardment. An intense period of bombardment of the Earth by meteorites around 4000 million years ago.

H.I and H.II regions. Clouds of hydrogen in the Galaxy. In H.I regions the hydrogen is neutral; in H.II regions the hydrogen is ionized, and the presence of hot stars will make the cloud shine as a nebula.

HABITABLE zone. See Goldilocks zone.

HALO, galactic. The spherical-shaped cloud of stars round the main part of the Galaxy.

HELIOSPHERE. The area round the Sun extending to between 50 and 100 a.u. where the Sun's influence is dominant. The boundary, where the solar wind merges with the interstellar medium, is called the heliopause.

HERTZSPRUNG–RUSSELL diagram (usually known as the H–R Diagram). A diagram in which stars are plotted according to the spectral types and their absolute magnitudes.

HIGH-VELOCITY star. A star travelling at more than around 40 miles per second (65 kilometres per second) in relation to the Sun. These are old stars, which do not share the Sun's motion round the galactic centre, but travel in more elliptical orbits.

HORIZON. The great circle on the celestial sphere which is everywhere 90 degrees from the observer's zenith.

HOT Jupiter. A class of exoplanet whose mass is close to or exceeds that of Jupiter and which orbits their parent stars at a distance of less than 1 Astronomical Unit.

HUBBLE constant. A measure of the rate at which a galaxy is receding. The current value is about 70 kilometres per second per megaparsec.

INFERIOR planets. Mercury and Venus, whose distances from the Sun are less than that of the Earth.

INFRARED radiation. Radiation with wavelengths longer than that of visible light, approximately above 700 nanometres.

INTERFEROMETER, stellar. An instrument for measuring star diameters. The principle is based upon light interference.

IONOSPHERE. The region of the Earth's atmosphere lying above the stratosphere.

KELVIN scale. A scale of temperature. 1 K is equal to 1 degree Celsius, but the Kelvin scale starts at absolute zero (–273.16°C).

KEPLER'S laws of planetary motion. These were laid down by Johannes Kepler, from 1609 to 1618. They are: (1) The planets move in elliptical orbits, with the Sun occupying one focus. (2) The imaginary line joining the centre of the planet to the centre of the Sun, sweeps out equal areas in equal times. (3) With a planet, the square of the sidereal period is proportional to the cube of the mean distance from the Sun.

KIRKWOOD gaps. Gaps in the main asteroid belt, where the periods would be commensurate with that of Jupiter – so that Jupiter keeps these areas 'swept clear'.

MAIN Sequence. The part of the Hertzsprung-Russell diagram where dwarf stars are burning hydrogen to helium. Most of the stars in the Universe are on the Main Sequence.

MERIDIAN. An imaginary circle passing through the north and south celestical poles, and through the observer's zenith (the point directly overhead). Prime meridian passes through the 0 degrees longitude at Greenwich, London, UK.

NANOMETRE. One thousand-millionth of a metre.

NEBULA. A cloud of gas and dust in interstellar space.

NEUTRINO. Small, lightweight particles that are by-products of nuclear fusion.

NEUTRON. Subatomic particle without an electric charge, one of the two constituents of atomic nuclei – the other being the proton. Neutrons and protons are approximately the same weight.

NEUTRON star. The remnant of a massive star which has exploded as a supernova. Neutron stars send out rapidly varying radio emissions, and are therefore called 'pulsars'. Only two (the Crab and Vela pulsars) have as yet been identified with optical objects.

NOVA. A star which suddenly flares up to many times its normal brilliancy, remaining bright for a relatively short time before fading back to obscurity.

OCCULTATION. The covering-up of one celestial body by another.

OORT cloud. An assumed spherical shell of comets surrounding the Solar system, at a range of around one light-year.

OPPOSITION. The position of a planet when exactly opposite to the Sun in the sky; the Sun, the Earth and the planet are then approximately lined up.

ORBIT. The path of a celestial object.

PARALLAX, trigonometrical. The apparent shift of an object when observed from two different directions.

PARSEC. The distance at which a star would have a parallax of one second of arc: 3.26 light-years, 206,265 astronomical units, or

30.857 million million kilometres.

PENUMBRA. (1) The area of partial shadow to either side of the main cone of shadow cast by the Earth. (2) The lighter part of a sunspot.

PERIGEE. The position of the Moon in its orbit when closest to the Earth.

PERIHELION. The position in orbit of a planet or other body when closest to the Sun.

PERTURBATIONS. The disturbances in the orbit of a celestial body produced by the gravitational effects of other bodies.

PHASES. The apparent changes in shape of the Moon and the inferior planets from new to full. Mars may show a gibbous phase, but with the other planets there are no appreciable phases as seen from Earth.

PHOTOMETER. An instrument used to measure the intensity of light from any particular source.

PHOTOMETRY. The measurement of the intensity of light.

PHOTON. The smallest 'unit' of light.

PHOTOSPHERE. The bright surface of the Sun.

PLANETARY nebula. A small, dense, hot star surrounded by a shell of gas. Planetary nebula are neither planets nor nebulae.

PLANETOID. Obsolete name for an asteroid (minor planet).

PLASMA. An ionized gas: a mixture of electrons and atomic nuclei.

POLES, celestial. The north and south points of the celestial sphere.

POPULATIONS, stellar. Two main types of star regions: I (in which the brightest stars are hot and bluish), and II (in which the brightest stars are old Red Giants.

PRECESSION. The apparent slow movement of the celestial poles. This also means a shift of the celestial equator, and hence of the equinoxes; the vernal equinox moves by 50 seconds of arc yearly, and has moved out of Aries into Pisces. Precession is due to the pull of the Moon and Sun on the Earth's equatorial bulge.

PRIME Meridian. The meridian on the Earth's surface which passes through the Airy Transit Circle at Greenwich Observatory. It is taken as longitude 0°.

PROMINENCES. Masses of glowing gas rising from the surface of the Sun. They are made up chiefly of hydrogen.

PROPER motion, stellar. The individual movement of a star on the celestial sphere.

PROTON. A positively charged subatomic particle, the main constituent of the atomic nuleus along with the neutron. It is approximately the same weight as the neutron.

PROTOPLANET. A body forming by the accretion of material, which will ultimately develop into a planet.

PROTOSTAR. The earliest stage in the formation of a star.

PULSAR. A rotating neutron star, often a strong radio source. Not all pulsars can be detected by radio, since the radiation is emitted in beams, and it depends upon whether these beams sweep over the Earth.

QUANTUM. The amount of energy possessed by one photon of light.

QUASAR. The core of a very powerful, remote active galaxy. The term QSO (quasi-stellar object) is also used.

RADIAL velocity. The movement of a celestial body toward or away from the observer; positive if receding, negative if approaching.

RADIANT. The point in the sky from which the meteors of any particular shower seem to radiate.

REGOLITH. A layer of loose rock and mineral grains on the surface of a planetary body. With the addition of organic material it becomes a soil.

RETROGRADE motion. Orbital or rotational movement opposite to that of the Earth's motion.

REVERSING layer. The gaseous layer above the Sun's photosphere.

RIGHT ascension. The angular distance of a celestial body from the vernal equinox, measured eastward. It is usually given in hours, minutes and seconds of time, so that the right ascension is the time difference between the culmination of the vernal equinox and the culmination of the body.

ROCHE limit. The distance from the centre of a planet within which a second body would be broken up by the planet's gravitational pull. Note, however, that this would be the case only for a body which had no appreciable gravitational cohesion.

SCHWARZSCHILD radius. The radius that a body must have if its escape velocity is to be equal to the velocity of light.

SCINTILLATION. Twinkling of a star due to the Earth's atmosphere. Planets may also show scintillation when low in the sky.

SELENOGRAPHY. The study of the Moon's surface.

SELENOLOGY. The lunar equivalent of geology.

SEYFERT galaxies. Galaxies with relatively small, bright nuclei and weak spiral arms. Some of them are strong radio emitters.

SIDEREAL period. The revolution period of a planet round the Sun, or of a satellite round its primary planet.

SIDEREAL time. The local time reckoned according to the apparent rotation of the celestial sphere. When the vernal equinox crosses the observer's meridian, the sidereal time is 0 hours.

SOLAR nebula. The cloud of interstellar gas and dust from which the Solar System was formed – around 5000 million years ago.

SOLAR wind. A flow of atomic particles streaming out constantly from the Sun in all directions.

SPECTROHELIOGRAPH. An instrument used for photographing the Sun in the light of one particular wavelength only. The visual equivalent of the spectroheliograph is the spectrohelioscope.

SPECTROSCOPIC binary. A binary system whose components are too close together to be seen individually, but which can be studied by means of spectroscopic analysis.

STARBURST galaxy. A galaxy in which there is an exceptionally high rate of star formation.

STELLAR wind. A continuous outflow of particles from a star, resulting in loss of mass.

SUPERIOR planets. All the planets lying beyond the orbit of the Earth in the Solar System (that is to say, all the principal planets apart from Mercury and Venus).

SUPERLUMINAL motion. The apparent movement of material at a velocity greater than that of light. It is purely a geometrical effect.

SUPERNOVA. A colossal stellar outburst, involving (1) the total destruction of the white dwarf member of a binary system, or (2) the collapse of a very massive star.

SYNCHRONOUS rotation. If the rotation period of a planetary body is equal to its orbital period, the rotation is said to be synchronous (or captured). The Moon and most planetary satellites have synchronous rotation.

T-TAURI stars. Cool, young stars of low to intermediate mass.

TEKTITES. Small, glassy objects found in a few localized areas of the Earth. They are not now believed to be meteoritic.

TERMINATOR. The boundary between the day and night hemispheres of the Moon or a planet.

TRANSIT. (1) The passage of a celestial body across the observer's meridian. (2) The projection of Mercury or Venus against the face of the Sun.

TROPOSPHERE. The lowest part of the Earth's atmosphere; its top lies at an average height of about 11 km. Above it lies the stratosphere; and above the stratosphere are the ionosphere and the exosphere.

TWILIGHT. The state of illumination when the sun is below the horizon by less than 18 degrees.

UMBRA. (1) The main cone of shadow cast by the Earth. (2) The darkest part of a sunspot.

VAN Allen zones. Zones of charged particles around the Earth. There are two main zones; the outer (made up chiefly of electrons) and the inner (made up chiefly of protons).

VARIABLE stars. Stars which change in brilliancy over short periods. They are of various types.

WHITE dwarf. A very small, very dense star which has used up its nuclear energy, and is in a very late stage of its evolution.

WOLF-RAYET stars. Very hot, greenish-white stars which are surrounded by expanding gaseous envelopes. Their spectra show bright (emission) lines.

YEAR. (1) Sidereal: the period taken for the Earth to complete one journey round the Sun (365.26 days). (2) Tropical: the interval between successive passages of the Sun across the vernal equinox (365.24 days). (3) Anomalistic: the interval between successive perihelion passages of the Earth (365.26 days; slightly less than 5 minutes longer than the sidereal year, because the position of the perihelion point moves along the Earth's orbit by about 11 seconds of arc every year). (4) Calendar: the mean length of the year according to the Gregorian calendar (365.24 days, or 365 days 5 hours 49 minutes 12 seconds).

ZENITH. The observer's overhead point (altitude 90°).

ZENITH distance. The angular distance of a celestial object from the Zenith.

ZODIAC. A belt stretching round the sky, 8° to either side of the ecliptic, in which the Sun, Moon and principal planets are to be found at any time. (Pluto is the only planet which can leave the Zodiac, though many asteroids do so.)

ZODIACAL Dust. The dust found in our Solar System which reflects the Zodiacal Light.

ZODIACAL Light. A cone of light rising from the horizon and stretching along the ecliptic; visible only when the Sun is a little way below the horizon. It is due to thinly-spread interplanetary material near the main plane of the Solar System.

ZONE of avoidance. The sky region near the plane of the Milky Way in which few galaxies can be seen, because of obscuration by interstellar dust in our Galaxy.

CREDITS

4-5 NASA

6-7 NASA/JPL/Caltech

11 James Symonds

14 t ESA/NASA m NASA b NASA

15 t/b NASA

16 t ISS/NASA b NASA

17 NASA

18 t/b NASA

19 t/b NASA

20 t/b NASA

21 Lick Observatory

22 t/b NASA

23 t NASA/GSFC/ASU/Noah Petro b NASA

24 t NASA/Alan Friedman m NASA b Damian Peach

25 t NASA b NASA

26 t NASA/GSFC/ASU/LRO m NASA b NASA/US Geological Survey

27 t NASA b NASA/ESA

28 t ESO b Pete Lawrence

29 t NASA/GSFC/ASU b NASA/GSFC/ASU

30 t NASA b Hinode/JAXA/NASA

30–31 background High Altitude Observatory/National Center for Atmospheric Research/SOHO/NASA/James Symonds

31 b Pete Lawrence

32–33 t Trace Project/Stanford Lockheed Institute for Space Research/NASA b Solar Dynamics Observatory/NASA/GSFC

34 t NASA/ESA/SOHO b Big Bear Solar Observatory.NJIT

35 t NASA m NASA b NASA

36–37 James Symonds r SOHO/NASA

38–39 SOHO/NASA

40 t SOHO/NASA b SOHO/NASA

41 NASA

42 t NASA/Johns Hopkins University Applied Physics Laboratory/Carnegie Institution of Washington b NASA/Johns Hopkins University Applied Physics Laboratory/Carnegie Institution of Washington

43 t NASA/Johns Hopkins University Applied Physics Laboratory/Carnegie Institution of Washington bl NASA/Johns Hopkins University Applied Physics Laboratory/Carnegie Institution of Washington br NASA

44 t NASA/Johns Hopkins University Applied Physics Laboratory/Carnegie Institution of Washington b NASA

45 t NASA/JPL b NASA/Johns Hopkins University Applied Physics Laboratory/Carnegie Institution of Washington

46 NASA/JPL

47 Magellan/NASA/JPL

48 t NASA/JPL/ESA b Magellan/NASA/JPL

49 t NASA/JPL b Magellan/NASA/JPL

50 NASA/James Symonds

51 t ESO b Brian May/James Symonds

52 t NASA m ESO b Kuiper Airborne Observatory/NASA

53 t UH/IA m Leonid Kulik b ISAS/JAXA

54 tl NASA/JPL b NASA/JPL/University of Arizona
55 t NASA/JPL/University of Arizona b NASA/JPL/University of Arizona
56 t NASA/JPL/CalTech b ESA/DLR/FU Berlin
57 t NASA b HiRise/MRO/JPL/NASA
58 t NASA/JPL/University of Arizona b NASA
59 t ESA/DLR/FU Berlin b NASA/JPL
60 NASA
61 t Mars Global Surveyor/MSSS/JPL/NASA m NASA/MOLA b Mars Global Surveyor/MSSS/JPL/NASA
62 t NASA/JPL/University of Arizona m HiRise/MRO/JPL/NASA b NASA/JPL-Caltech/University of Arizona/Texas A&M University
63 t NASA/JPL-Caltech/University of Arizona/Texas A&M University m NASA/JPL-Caltech/University of Arizona/Texas A&M University b NASA/JPL-Caltech/University of Arizona/Texas A&M University
64 t NASA b NASA/JPL/Caltech/Cornell University
65 t NASA/JPL/Cornell m NASA/JPL/Cornell b NASA/JPL/Caltech/Cornell University
66-67 NASA/JPL-Caltech/Cornell University
68 t/m/b NASA/JPL/University of Arizona
69 t/b NASA
70 t NASA/JPL-Caltech/UCLA/MPS/DLR/DA m/b NASA/JPL-Caltech/UCLA/MPS/DLR/IDA
71 NASA /JPL/ Caltech; Justin Cowart, The Planetary Society
72 t/m NASA b ESA/NASA
73 James Symonds
74 t H. Hammel, MIT and NASA/ESA bl (NASA/SWRI/R. Gladstone et al./HST/J. Clarke et al./R. Beebe et al. br NASA
75 t John T. Clarke, University of Michigan bl/br NASA
76 t NASA b NASA/A. Simon-Millar (NASA/GSFC)/I. de Pater and M Wong, University of California, Berkeley
77 NASA/JPL
78 t/b NASA/JPL
79 t NASA/JPL b NASA/JPL/USGS
80 t R. Pappalardo/Galileo Project/ NASA/JPL b NASA
81 t Galileo/NASA/JPL bl Karkoschka/NASA br Galileo/NASA/JPL
82 t NASA/JPL/Caltech bl NASA/JPL/University of Colorado br NASA/JPL
83 t/m/ Cassini/NASA/JPL
84-85 Cassini Imaging Team/ISS/JPL/ESA/NASA
86 t/m Cassini/NASA/JPL b NASA/JPL/University of Arizona
87 t Cassini/NASA/JPL b Karkoschka/NASA
88 t/b Cassini/SSI/NASA/ESA/JPL
89 t/m/b Cassini/NASA/JPL
90 NASA/JPL/STI
91 t/b NASA/JPL
92 t NASA/STI b NASA

93 tl NASA/HST tr NASA b Lawrence Sromovsky, University of Wisconsin-Madison/W. M. Keck Observatory
94 t NASA/JPL bl Lawrence Sromovsky, University of Wisconsin-Madison, NASA br Keck
95 NASA/JPL
96 NASA
97 t NASA m ESA/ESO/NASA b NASA /Johns Hopkins University Applied Physics Laboratory / Southwest Research Institute
98 t NASA/ESA/M. Brown, Caltech b NASA/JPL/Caltech
99 t Palomar b ESA
100 t/m/b NASA/JPL
101 t NASA b NASA/JPL
102-103 James Symonds
104-105 NASA, ESA, Digitized Sky Survey 2/Davide De Martin, ESA/Hubble
106 t DSS/UKSchmidt/StSci m NASACXC/SAO b UK Schmidt/AAO
107 t Nik Szymanek/Ian King m Pete Lawrence b ESO
108 NASA/JPL/Caltech
109 t ESA b ESO
110 NASA/ESA/P. Kalas, J. Graham, E. Chiang, E. Kite, University California, Berkeley, M. Clampin, NASA/Goddard/M. Fitzgerald, Lawrence Livermore / K. Stapelfeldt, J. Krist, NASA/JPL
111 Anglo-Australian Observatory, Digitized Sky Survey, Davide de Martin
112 ESA/XMM-Newton/EPIC
113 t 2MASS/NASA b Greg Parker
114 Greg Parker
115 t R.Hurt/SSC-Caltech/JPL-Caltech/NASA b NASA
116 t Pete Lawrence b NASA/JPL/GALEX/Caltech/OCIW
117 t Greg Parker b ESO
118-119 HST/NASA
120 t NASA b NASA/James Symonds
121 t NASA b NASA/James Symonds
122 t ESO/Beletsky
123 ESO/SFRC
124 t NASA/HST bl Greg Parker br ESO
125 t Pete Lawrence b ESO/Digitized Sky Survey 2/Davide De Martin
126 NASA/JPL/Caltech/Iowa State University
127 Hubble Heritage Team/A. Riess, STScI/NASA
128 NASA/JPL/M. Marengo/Iowa State University
129 NASA/STScI/Noel Carboni
131 NASA/AMES/JPL/Caltech
132 ESO/J. Emerson/Vista
133 NASA/AMES/JPL-Caltech/STScI
134 l ESO/J. Emerson/Vista r NASA/JPL/Caltech/STScI
135 r Greg Parker

136 Daniel Cantin, McGill University
137 Ian Morrison/Jodrell Bank/University of Manchester
138 Spitzer Space Telescope/NASA/JPL-Caltech/Harvard-Smithsonian CfA
139 HST/NASA/STScI
140 t HST/NASA/STScI b HST/NASA/STScI
141 HST/NASA/STScI
142-143 HST/CXCSAO/NASA/STScI
144 bl ESO r ESO
145 t HST/NASA/STScI r Spitzer Space Telescope/NASA/JPL-Caltech/Harvard-Smithsonian CfA
146 t/b T. A. Rector, B. A. Wolpa/NOAO/AURA
147 NASA
148 ESO
149 t HST/NASA/STScI b ESO
150-151 NASA/ESA/Hubble SM4 ERO Team t ESO
152 NRAO
153 t NASA/ESA/H.E. Bond/STScI
154 NASAE/SA/SSC/CXC/STScI
155 Hubble: NASA/ESA/D. Q. Wang, University of. Massachusetts, Amherst/Spitzer Space Telescope/NASA,/JPL/S. Stolovy, SSC/Caltech
156 ESO/S.Brunier
156 l Nik Szymanek r ESO/S. Brunier
158 t NASA/ESA/M. Livio/STScI b AURA/NOAO/NSF
159 NASA/JOL/Caltech/STScI
160 t/m HST b ESO
161 t NASA/JPL/Caltech/STScI b ESO
162 NASA/JPL/Caltech/STScI
163 Subaru/NAOJ
164 ESO
165 Greg Parker
166 t J.-C. Cuillandre, CFHT/CFHT b Spitzer Space Telescope/NASA
167 t Spitzer Space Telescope/NASA bNASA/HST
168 NASA/HST/STScI
169 t NASA/ESA/Hubble Heritage Team/STScI/AURA b NRAO/NSF b Brad Whitmore, NASA b ALMA/ESO/NAOJ/NRAO
170 t B. Whitmore NASA/STScI b X-ray: NASA/CXC/SAO/J.DePasquale; IR: NASA/JPL-Caltech; Optical: NASA/STScI
171 t J. A. Biretta et al., Hubble Heritage Team, STScI/AURA/NASA b NASA,/National Radio Astronomy Observatory/National Science Foundation/John Biretta, STScI/JHU, Associated Universities, Inc.
173 NASA/ESA/R. Williams/HUDF team
174 NASA/ESA/William Keel/Hanny van Arkel
175 t ESA/NASA/J.-P. Kneib/R. Ellis, Caltech bl NASA/ESA br J. Rhoads, STScI/WIYN/AURA/NOAO/NSF
176 NASA/ESA/STScI
177 WMAP Science Team/NASA

INDEX

THE AUTHORS

Brian May, PhD, CBE, ARCS, FRAS, Chancellor of Liverpool John Moores University (UK) is a founding member of the rock group Queen, a world-renowned guitarist, songwriter, producer and performer. A producer and musical director of the award-winning Queen musical We Will Rock You, in 2012 he celebrated 10 years of performance to packed houses in London's Dominion Theatre. He has published research articles in the field of the solar zodiacal dust cloud, is co-author, with Sir Patrick Moore and Chris Lintott, of Bang! The Complete History of the Universe, and, as a lifelong devotee of 3D imaging he wrote, with photohistorian Elena Vidal, A Village Lost and Found, a book bringing the important work of 19th-century stereoscopic photographer T. R. Williams into the 21st century. Brian is a regular contributor to BBC TV's The Sky at Night. He is patron to a number of charities including The Mercury Phoenix Trust and The British Bone Marrow Donor Association. Founder of the Save-Me campaign, he is a passionate campaigner for the welfare of wild animals, and the abolition of blood sports. Brian likes to interact with his fans who can contact him and enjoy updates on his work and thoughts via his website at www.brianmay.com.

Sir Patrick Moore CBE, FRS, FRAS specialized in the study of the Moon, and was the leading interpreter of modern astronomy, inspiring several generations of astronomers to a lifelong pursuit of the subject. His monthly BBC TV show, The Sky at Night, was launched in 1957, and celebrated its 55th year in 2012, a unique broadcasting record. He explained newsworthy events in space and astronomy for the BBC throughout this entire period, making countless appearances on topical TV and radio shows. He wrote hundreds of books and articles, and lectured to packed venues the world over. He was a Fellow of the Royal Society, an Honorary Vice President of the Royal Astronomical Society, an Honorary Vice President of the British Astronomical Association, a member of the International Astronomy Union, an Honorary Vice President of the Society for the History of Astronomy, and was awarded 12 honorary degrees from British Universities. An accomplished musician, he has over 100 compositions to his name, including a march for the Band of the Royal Marines. Moore was a campaigner against fox-hunting and an active supporter of several major charities. He shared his home with his beloved cat Ptolemy, who appeared on the The Sky at Night.

Chris Lintott PhD, FRAS is a researcher in the Department of Physics at the University of Oxford, where he is also a junior research fellow at New College. His research focuses on galaxy formation and evolution, primarily carried out in collaboration with the hundreds of thousands of volunteers who take part in Galaxy Zoo and other citizen science projects at Zooniverse.org. He is the chair of the transatlantic Citizen Science Alliance, which aims to develop projects that allow the public to make real contributions to research, and received the 2011 Royal Society Kohn Award for this work. He is best-known as co-presenter of the BBC's long-running Sky at Night, having first appeared on the show as a callow youth back in 2000. Away from astronomy, he's likely to be found enjoying a glass of wine, cooking, going to the opera or supporting his twin loves, Torquay United and the Chicago Fire.